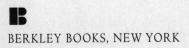

BERKLEY BOOKS, NEW YORK

David Bodanis

A Biography
of the World's Most
Famous Equation

$E=mc^2$

A Berkley Book
Published by The Berkley Publishing Group
A division of Penguin Putnam Inc.
375 Hudson Street
New York, New York 10014

Copyright © 2000 by David Bodanis.
Text design by Ralph L. Fowler.

Photograph of Einstein on title page from the Leo Baeck Institute / Archive Photos.

Published by arrangement with Walker Publishing Company, Inc.

PRINTING HISTORY
Walker & Company hardcover edition / September 2000
Berkley trade paperback edition / October 2001

Berkley trade paperback ISBN: 0-425-18164-2

Library of Congress Cataloging-in-Publication Data

Bodanis, David.
    E=mc² : a biography of the world's most famous equation / David Bodanis.
        p. cm.
    Originally published: New York : Walker, 2000.
    Includes bibliographical references and index.
    ISBN 0-425-18164-2
    1. Force and energy. 2. Mass (Physics) 3. Mathematical physics. 4. Einstein, Albert, 1879–1955. I. Title.

QC73.8.C6 B63 2001
530.11—dc21

                                                                    2001046083

PRINTED IN THE UNITED STATES OF AMERICA

Contents

$$E=mc^2$$

Preface   vii

PART 1 Birth

   1   **Bern Patent Office, 1905**   3

PART 2 Ancestors of $E=mc^2$

   2   **E Is for Energy**   11

   3   **=**   23

   4   **m Is for mass**   27

   5   **c Is for celeritas**   37

   6   **²**   55

PART 3 The Early Years

   7   **Einstein and the Equation**   73

   8   **Into the Atom**   93

   9   **Quiet in the Midday Snow**   100

CONTENTS

PART 4 Adulthood

10 **Germany's Turn** 117

11 **Norway** 134

12 **America's Turn** 143

13 **8:16 A.M. — Over Japan** 163

PART 5 Till the End of Time

14 **The Fires of the Sun** 173

15 **Creating the Earth** 184

16 **A Brahmin Lifts His Eyes Unto the Sky** 195

Epilogue:
   What Else Einstein Did 204
Appendix:
   Follow-Up of Other Key Participants 221
Notes 237
Guide to Further Reading 301
Acknowledgments 319
Index 325

$$E=mc^2$$

A while ago I was reading an interview with the actress Cameron Diaz in a movie magazine. At the end the interviewer asked her if there was anything she wanted to know, and she said she'd like to know what $E=mc^2$ really means. They both laughed, then Diaz mumbled that she'd meant it, and then the interview ended.

"You think she did mean it?" one of my friends asked, after I read it aloud. I shrugged, but everyone else in the room—architects, two programmers, and even one historian (my wife!)—was adamant. They knew exactly what she intended: They wouldn't mind understanding what the famous equation meant too.

It got me thinking. Everyone knows that $E=mc^2$ is really important, but they usually don't know what it means, and that's frustrating, because the equation is so short that you'd think it would be understandable.

There are plenty of books that try to explain it, but who can honestly say they understand them? To most readers they contain just a mass of odd diagrams—those little trains or rocketships or flashlights that are utterly mystifying. Even firsthand instruction doesn't always help, as Chaim Weizmann found when he took a long

Atlantic crossing with Einstein in 1921: "Einstein explained his theory to me every day," Weizmann said, "and soon I was fully convinced that he understood it."

I realized there could be a different approach. The overall surveys of relativity fail not because they're poorly written, but because they take on too much. Instead of writing yet another account of all of relativity, let alone another biography of Einstein—those are interesting topics, but have been done to death—I could simply write about E=mc². That's possible, for it's just one part of Einstein's wider work. To a large extent, it stands on its own.

The moment I started thinking this way, it became clear how to go ahead. Instead of using the rocketship-and-flashlight approach, I could write the biography of E=mc². Everyone knows that a biography entails stories of the ancestors, childhood, adolescence, and adulthood of your subject. It's the same with the equation.

The book begins, accordingly, with the history of each part of the equation—the symbols E, m, c, =, and ². For each of these—the equation's "ancestors"—I focus on a single person or research group whose work was especially important in creating our modern understanding of the terms.

Once the nature of those symbols is clear, it's time to turn to the equation's "birth." This is where Einstein enters the book: his life as a patent clerk in 1905; what he'd been reading, and what he'd been thinking about, which led to all those symbols he wove together in the equation hurtling into place in his mind.

If the equation and its operations had stayed solely in Einstein's hands, our book would simply have continued with Einstein's life after 1905. But pretty quickly after this great discovery his interests shifted to other topics; his personal story fades from the book, and instead we pick up with other physicists: more empirical ones now, such as the booming, rugby-playing Ernest Rutherford, and the quiet, ex-POW James Chadwick, who together helped reveal the detailed structures within the atom that could—in principle—be manipulated to allow the great power the equation spoke of to come out.

In any other century those theoretical discoveries might have taken a long time to be turned into practical reality, but the details of how Einstein's equation might be used became clear early in 1939, just as the twentieth century's greatest war was beginning. A long, central section of the book homes in on the equation's coming of age here, in the furious race between scientists based in the United States and those in Nazi Germany to see who could build a deathly, planet-controlling bomb first. The history is often presented as if America's victory were inevitable, due to the country's industrial superiority, but it turns out that Germany came dangerously closer to success than is often realized. Even as late as D Day in June 1944, Army Chief of Staff George Marshall saw to it that several of the U.S. units landing in France were supplied with Geiger counters as a precaution against a possible Nazi attack with radioactive weapons.

In the final sections of the book we switch away from war; the equation's "adulthood" has begun. We'll see how $E=mc^2$ is at the heart of many medical devices, such as the PET scanners used for finding tumors; its effects are also widespread in our ordinary household devices, including televisions and smoke alarms. But even more significant is how its power stretches far out into the universe, helping to explain how stars ignite, and our planet keeps warm; how black holes are created, and how our world will end. At the very end of the book, there are detailed notes, for readers interested in more mathematical or historical depth; further background explanations are available at the Web site dbodanis.com.

The stories along the way are as much about passion, love, and revenge as they are about cool scientific discovery. There will be Michael Faraday, a boy from a poor London family desperate for a mentor to lift him to a better life, and Emilie du Châtelet, a woman trapped in the wrong century, trying to carve out a space where she wouldn't be mocked for using her mind. There are accounts of Knut Haukelid and a team of fellow young Norwegians, forced to attack their own countrymen to avert a greater Nazi evil; Cecilia Payne, an Englishwoman who finds her career destroyed after daring to glimpse the sun's fate in the year A.D. 6 billion; and a nineteen-year-old Brahmin, Subrahmanyan Chandrasekhar, who discovers something even more fearful, out in the beating heat of the Arabian Sea in midsummer. Through all their stories—as well as highlights from those of Isaac Newton, Werner Heisenberg, and other researchers—the meaning of each part of the equation becomes clear.

PART I

Birth

I

$$E=mc^2$$

From THE COLLECTED PAPERS OF ALBERT EINSTEIN, VOLUME I:

13 April 1901

Professor Wilhelm Ostwald
University of Leipzig
Leipzig, Germany

Esteemed Herr Professor!

Please forgive a father who is so bold as to turn to you, esteemed Herr Professor, in the interest of his son.

I shall start by telling you that my son Albert is 22 years old, that . . . he feels profoundly unhappy with his present lack of position, and his idea that he has gone off the tracks with his career & is now out of touch gets more and more entrenched each day. In addition, he is oppressed by the thought that he is a burden on us, people of modest means. . . .

I have taken the liberty of turning [to you] with the humble request to . . . write him, if possible, a few words of encouragement, so that he might recover his joy in living and working.

If, in addition, you could secure him an Assistant's position for now or the next autumn, my gratitude would know no bounds. . . .

I am also taking the liberty of mentioning that my son does not know anything about my unusual step.

I remain, highly esteemed Herr Professor, your devoted

Hermann Einstein

No answer from Professor Ostwald was ever received.

The world of 1905 seems distant to us now, but there were many similarities to life today. European newspapers complained that there were too many American tourists, while Americans were complaining that there were too many immigrants. The older generation everywhere complained that the young were disrespectful, while politicians in Europe and America worried about the disturbing turbulence in Russia. There were newfangled "aerobics" classes; there was a trend-setting vegetarian society, and calls for sexual freedom (which were rebuffed by traditionalists standing for family values), and much else.

The year 1905 was also when Einstein wrote a series of papers that changed our view of the universe forever. On the surface, he seemed to have been leading a pleasant, quiet life until then. He had often been interested in physics puzzles as a child, and was now a recent university graduate, easygoing enough to have many friends. He had married a bright fellow student, Mileva, and was earning enough money from a civil service job in the patent office to spend his evenings and Sundays in pub visits, or long walks—above all, he had a great deal of time to think.

Although his father's letter hadn't succeeded, a friend of Einstein's from the university, Marcel Grossman, had pulled the right strings to get Einstein the patent job in 1902. Grossman's help was necessary not so much because Einstein's final university grades were unusually low—through cramming with the ever-useful Grossman's notes, Einstein had just managed to reach a 4.91 average out of a possible 6, which was almost average—but because one professor, furious at Einstein for telling jokes and cutting classes, had spitefully written unacceptable references. Teachers over the years had been irritated by his lack of obedience, most notably Einstein's high school Greek grammar teacher, Joseph Degenhart, the one who has achieved immortality in the history books through insisting that "nothing would ever become of you." Later, when told it would be best if he left the school, Degenhart had explained, "Your presence in the class destroys the respect of the students."

Outwardly Einstein appeared confident, and would joke with his friends about the way everyone in authority seemed to enjoy putting him down. The year before, in 1904, he had applied for a promotion from patent clerk third class to patent clerk second class. His supervisor, Dr. Haller, had rejected him, writing in an assessment that although Einstein had "displayed some quite good achievements," he would still have to wait "until he has become fully familiar with mechanical engineering."

In reality, though, the lack of success was becoming serious. Einstein and his wife had given away their first child, a daughter born before they were married, and were now trying to raise the second on a patent clerk's salary. Einstein was twenty-six. He couldn't even afford the money for part-time help to let his wife go back to

Albert Einstein at his high school graduation in Aarau, Switzerland
ETH, ZURICH

her studies. Was he really as wise as his adoring younger sister, Maja, had told him?

He managed to get a few physics articles published, but they weren't especially impressive. He was always aiming for grand linkages—his very first paper, published back in 1901, had tried to show that the forces controlling the way liquid rises up in a drinking straw were similar, fundamentally, to Newton's laws of gravitation. But he could not quite manage to get these great linkages to work, and he got almost no response from other physicists. He wrote to his sister, wondering if he'd ever make it.

Even the hours he had to keep at the patent office worked against him. By the time he got off for the day, the one science library in Bern was usually closed. How would he have a chance if he couldn't even stay up to date with the latest findings? When he did have a few free moments during the day, he would scribble on sheets he kept in one drawer of his desk—which he jokingly called his department of theoretical physics. But Haller kept a strict eye on him, and the drawer stayed closed most of the time. Einstein was slipping behind, measurably, compared to the friends he'd made at the university. He talked with his wife about quitting Bern and trying to find a job teaching high school. But even that wasn't any guarantee: he had tried it before, only four years earlier, but never managed to get a permanent post.

And then, on what Einstein later remembered as a beautiful day in the spring of 1905, he met his best friend, Michele Besso ("I like him a great deal," Einstein wrote, "because of his sharp mind and his simplicity"), for one of their long strolls on the outskirts of the city. Often they just gossiped about life at the patent office, and music, but today Einstein was uneasy. In the past few months a great deal of what he'd been thinking about had started coming together, but there was still something Einstein felt he was very near to understanding but couldn't quite see. That night Einstein still couldn't quite grasp it, but the next day he suddenly woke up, feeling "the greatest excitement."

It took just five or six weeks to write up a first draft of the article, filling thirty-some pages. It was the start of his theory of relativity. He sent the article to *Annalen der Physik* to be published, but a few weeks later, he realized that he had left something out. A three-page supplement was soon delivered to the same physics journal. He admitted to another friend that he was a little unsure how accurate the supplement was: "The idea is amusing and enticing, but whether the Lord is laughing at it and has played a trick on me—that I cannot know."

But in the text itself he began, confidently: "The results of an electrodynamic investigation recently published by me in this journal lead to a very interesting conclusion, which will be derived here." And then, four paragraphs from the end of this supplement, he wrote it out.

$E=mc^2$ had arrived in the world.

Ancestors of
E=mc²

2

$$E=mc^2$$

The word *energy* is surprisingly new, and can only be traced in its modern sense to the mid 1800s. It wasn't that people before then had not recognized that there were different powers around—the crackling of static electricity, or the billowing gust of a wind that snaps out a sail. It's just that they were thought of as unrelated things. There was no overarching notion of "Energy" within which all these diverse events could fit.

One of the men who took a central role in changing this was Michael Faraday, a very good apprentice bookbinder who had no interest, however, in spending his life binding books. As an escape hatch from poverty in the London of the 1810s, though, it was a job that had one singular advantage: "There were plenty of books there," he mused years later to a friend, "and I read them." But it was fragmentary reading, and Faraday recognized that, just snatching glimpses of pages as they came in to be bound. Occasionally he had evenings alone, next to the candles or lamps, reading longer sixteen- or thirty-two-page bound sheaves.

He might have stayed a bookbinder, but although social mobility in Georgian London was very low, it

Michael Faraday
ENGRAVING BY J. COCHRAN OF PAINTING BY H. W. PICKERGILL, ESQ., R.A.
AIP EMILIO SEGRÈ VISUAL ARCHIVES

wasn't quite zero. When Faraday was twenty, a shop visitor offered him tickets to a series of lectures at the Royal Institution. Sir Humphry Davy was speaking on electricity, and on the hidden powers that must exist behind the surface of our visible universe. Faraday went, and realized he had been granted a lucky glimpse of a better life than he had working at the shop. But how could he enter it? He had not been to Oxford, or to Cambridge or, indeed, even attended much of what we call secondary school; he had as much money as his blacksmith father could give him—none—and his friends were just as poor.

But he could bind an impressive-looking book. Faraday had always been in the habit of taking notes when he could, and he'd brought back to the shop notes he'd taken at Davy's lectures. He wrote them out, and inserted a few drawings of Davy's demonstration apparatus. Then he rewrote the manuscript—all his drafts are kept today with the attention due a sacred relic, in the basement Archive Room of London's Royal Institution—took up his leather, awls, and engraving tools, and bound together a terrific book, which he sent to Sir Humphry Davy.

Davy replied that he wanted to meet Faraday. He liked him, and despite a disconcerting series of starts and stops, finally hired him away from the binder as a lab assistant.

Faraday's old shopmates might have been impressed, but his new position was not as ideal as he'd hoped. Sometimes Davy behaved as a warm mentor, but at other times, as Faraday wrote to his friends, Davy would seem angry, and push Faraday away. It was especially frustrating to Faraday, for he'd been drawn to science in large part by Davy's kind words; his hints that if one only had the skill and could see what had hitherto been hidden, everything we experience could actually be linked.

It took several years for Davy finally to ease up, and when he did it appeared to coincide with Faraday being asked to understand an extraordinary finding out of Denmark. Until then, everyone knew that electricity and magnetism were as unrelated as any two forces could be. Electricity was the crackling and hissing stuff that came from batteries. Magnetism was different, an invisible force that made navigators' needles tug forward, or pulled pieces of iron to a lodestone. Magnetism was not anything you thought of as part of batteries and circuits. Yet a lecturer in Copenhagen had now found that if you switched on the current in an electric wire, any compass needle put on top of the wire would turn slightly to the side.

No one could explain this. How could the power of electricity in a metal wire possibly leap out and make a magnetic compass needle turn? When Faraday, now in his late twenties, was asked to work on how this link might occur his letters immediately became more cheerful.

Sir Humphry Davy
PHOTO RESEARCHERS, INC.

He started courting a girl ("You know me as well or better than I do myself," he wrote. "You know my former prejudices, and my present thoughts—you know my weaknesses, my vanity, my whole mind") and the girl liked being courted: in mid-1821, when Faraday was twenty-nine, they got married. He became an official member of the church which his family had been a part of for many years. This was a gentle, literalist group called the Sandemanians, after Robert Sandeman, who'd brought the sect to England. Most of all, Faraday now had a chance to impress Sir Humphry: to pay him back for his initial faith in hiring a relatively uneducated young binder, and to cross, finally, the inexplicable barriers Davy had raised between them.

Faraday's limited formal education, curiously enough, turned out to be a great advantage. This doesn't happen often, because when a scientific subject reaches an advanced level, a lack of education usually makes it impossible for outsiders to get started. The doors are closed, the papers unreadable. But in these early days of understanding energy it was a different story. Most science students had been trained to show that any complicated motion could be broken down into a mix of pushes and pulls that worked in straight lines. It was natural for them, accordingly, to try to see if there were any straight-line pulls between magnets and electricity. But this approach didn't show how the power of electricity might tunnel through space to affect magnetism.

Because Faraday did not have that bias of thinking in straight lines, he could turn to the Bible for inspiration. The Sandemanian religious group he belonged to believed in a different geometric pattern: the circle. Humans are holy, they said, and we all owe an obligation to one another based on our holy nature. I will help you, and you will help the next person, and that person will help another, and so on until the circle is complete. This circle wasn't merely an abstract concept. Faraday had spent much of his free time for years either at the church talking about this circular relation, or engaged in charity and mutual helping to carry it out.

He got to work studying the relationship between electricity and magnetism in the late summer of 1821. It was twenty years before Alexander Graham Bell, the inventor of the telephone, would be born; more than fifty years before Einstein. Faraday propped up a magnet. From his religious background, he imagined a whirling tornado of invisible *circular* lines swirling around it. If he were right, then a loosely dangling wire could be tugged along, caught in those mystical circles like a small boat getting caught up in a whirlpool. He connected the battery.

And immediately he had the discovery of the century.

Later, the apocryphal story goes—after all the announcements, after Faraday was made a Fellow of the Royal Society—the prime minister of the day asked what good this invention could be, and Faraday answered: "Why, Prime Minister, someday you can tax it."

What Faraday had invented, in his basement laboratory, was the basis of the electric engine. A single dangled wire, whirling around and around, doesn't seem like much. But Faraday had only a small magnet, and was feeding in very little power. Rev it up, and that whirling wire will still doggedly follow the circular patterns he had mapped out in seemingly empty air. Ultimately one could attach heavy objects to a similar wire, and they would be tugged along as well—that's how an electric engine works. It doesn't matter whether it is the featherweight spinning plate of a computer drive that's being dragged along, or the pumps that pour tons of fuel into a jet engine.

Faraday's brother-in-law, George Barnard, remembered Faraday at the moment of discovery: "All at once he exclaimed, 'Do you see, do you see, do you see, George?' as the wire began to revolve. . . . I shall never forget the enthusiasm expressed in his face and the sparkling in his eyes!"

Faraday was sparkling because he was twenty-nine years old and had made a great discovery, and it really did seem to suggest that the deepest ideas of his religion were true. The crackling of electricity, and the silent force fields of a magnet—and now even the speeding motion of a fast twirling copper wire—were seen as linked. As the amount of electricity went up, the available magnetism would go down. Faraday's invisible whirling lines were the tunnel—the conduit—through which magnetism could pour into electricity, and vice versa. The full concept of "Energy" had still not been formed, but Faraday's discovery that these different kinds of energy were linked was bringing it closer.

It was the high point of Faraday's life—and then Sir Humphry Davy accused him of stealing the whole idea.

Davy began to let it be known that he had personally discussed the topic with a different researcher who had been investigating it—a properly educated researcher—and Faraday must have just overheard them.

The story was false, of course, and Faraday tried to protest, begging on the basis of their past friendship to let him explain, but Davy would have none of it. There were further crude hints, from others, if not from Davy himself: What else could you expect from a lower-class boy, from someone so junior, who was trying to wangle his way up as an apprentice; who knew nothing of what a more in-depth education could teach? After a few months Davy backed off, but he never apologized, and left the charges to dangle.

In notes and private journal entries Davy often wrote how important it was to encourage young men. The problem was that he just couldn't bring himself to do it. The issue was nothing as simple as youth versus old age. Davy was little more than a decade older than Faraday. But Davy loved being lionized as the leader of British science, and all the time he spent away from the lab, soaking up praise in London high society with his status-keen wife, meant that the praise was increasingly false. He wasn't really on top of the latest research. When he corresponded with thinkers on the Continent he knew they were impressed at getting a letter from someone so prominent in the Royal Institution, but he avoided offering fresh ideas.

Hardly anyone else recognized this, but Faraday did.

He was more like Davy than anyone else. Both men had started at a level much below that of their contemporaries in London science. Faraday made no excuses for that, but Davy did everything he could to hide his past. Faraday's quiet presence was a constant reminder of what they'd both once shared.

Faraday never spoke out against Davy. But for years after the charges of plagiarism and their repercussions, he stayed warily away from front-line research. Only when Davy died, in 1829, did he get back to work.

Faraday lived into old age, in time becoming prominent in the Royal Institution himself. His rise was typical of the move from gentlemanly to professional science. Davy's slurs against him were long forgotten. He went on to make other discoveries; he became very famous, and was often in demand, receiving such letters as this:

> May 28th, 1850
>
> Dear Sir,
>
> It has occurred to me that it would be extremely beneficial to a large class of the public to have some account of your late lectures on the breakfast-table. . . . I should be exceedingly glad to have . . . them published in my new enterprise. . . .
>
> With great respect and esteem I am Dear Sir,
>
> Your faithful servant,
>
> Charles Dickens

By the last decade of his life, though, Faraday—like Davy—was no longer able to follow the latest results. But the energy concept had taken on a life of its own. All the world's seemingly separate forces were slowly, majestically, being linked to create this masterpiece of the Victorian Age: the huge, unifying domain of Energy. Since Faraday had shown that even electricity and magnetism were linked—two items that had once seemed totally distinct—the scientific community was more confident that every other form of energy could similarly be shown to be deeply connected. There was chemical energy in an exploding gunpowder charge, and there was frictional heat energy in the scraping of your shoe, yet they were linked too. When a gunpowder charge went off, the amount of energy it produced in air blasts and falling rocks would be exactly the same as what had been in the resting chemical charge inside.

It's easy to miss how extraordinary a vision was the energy concept that Faraday's work helped create. It's as if when God created the universe, He had said, I'm going to put X amount of energy in this universe of mine. I will let stars grow and explode, and planets move in their orbits, and I will have people create great cities, and there will be battles that destroy those cities, and then I'll let the survivors create new civilizations. There will be fires and horses and oxen pulling carts; there will be coal and steam engines and factories and even mighty locomotives. Yet throughout the whole sequence, even though the types of energy that people see will change, even though sometimes the energy will appear as the heat of human or animal muscle, and sometimes it will appear as the gushing of waterfalls or the explosions of volcanoes: despite all those variations, the *total* amount of energy will remain the same. The amount I created at the beginning will not change. There will not be one millionth part less than what was there at the start.

Expressed like this it sounds like the sheerest mumbo jumbo—Faraday's religious vision of a single universe, with just one single force spreading all throughout it. It's something like Obi-Wan Kenobi's description in *Star Wars*: "The Force is the energy field created by all living things; it binds the galaxy together."

Yet it's true! When you swing closed a cupboard door, even if it's in the stillness of your home at night, energy will appear in the gliding movement of the door, but exactly that much energy was removed from your muscles. When the cupboard door finally closes, the energy of its movement won't disappear, but will simply be relocated to the shuddering bump of the door against the cupboard, and to the heat produced by the grinding friction of the hinge. If you had to dig your feet slightly against the floor to keep from slipping when closing the door, the earth will shift in its orbit and rebound upward by exactly the amount needed to balance that.

The balancing occurs everywhere. Measure the chemical energy in a big stack of unburned coal, then ignite it in a train's boiler and measure the energy of the roaring fire and the racing locomotive. Energy has clearly changed its forms; the systems look very different. But the total is exactly, precisely the same.

Faraday's work was part of the most successful program for further research the nineteenth century had seen. Every quantity in these energy transformations that Faraday and others had now unveiled could be computed and measured. When that was done, the results confirmed, always, that indeed the total sum had never changed—it was "conserved." This became known as the Law of the Conservation of Energy.

Everything was connected; everything neatly balanced. In the last decade of Faraday's life, Darwin seemed to have proven that God wasn't needed to create the living species on our planet. But Faraday's vision of an unchanging total Energy was often felt to be a satisfactory alternative: a proof that the hand of God really had touched our world, and was still active amid us.

This concept of energy conservation is what the science teachers in Einstein's cantonal high school in Aarau, in northern Switzerland, had taught him when he arrived there for remedial work in 1895, twenty-eight years after Faraday's death. Einstein had been sent to the school not because he'd had any desire to go there—he had already dropped out of one perfectly good high school in Germany, vowing that he'd had enough—but because he had failed his entrance exams at the Federal Institute of Technology in Zurich, the only university that offered a chance of taking a high school dropout. One kindly instructor there had thought he might have some merit, so instead of turning him away entirely, the institute's director had suggested this quiet school—set up on informal, student-centered lines—in the northern valleys.

When Einstein did finally make it into the Federal Institute of Technology—after his first delicious romance, with the eighteen-year-old daughter of his Aarau host—the physics lecturers there were still teaching the Victorian gospel, of a great overarching energy force. But Einstein felt his teachers had missed the point. They were not treating it as a live topic, honestly hunting for what it might mean, trying to feel for those background religious intimations that had driven Faraday and others forward. Instead, energy and its conservation was just a formalism to most of them, a set of rules. There was a great complacency throughout much of Western Europe at the time. European armies were the most powerful in the world; European ideas were "clearly" superior to those of all other civilizations. If Europe's top thinkers had concluded that energy conservation was true, then there was no reason to question them.

Einstein was easygoing about most things, but he couldn't bear complacency. He cut many of his college classes—teachers with that attitude weren't going to teach him anything. He was looking for something deeper, something broader. Faraday and the other Victorians had managed to widen the concept of energy until they felt it had encompassed every possible force.

But they were wrong.

Einstein didn't see it yet, but he was already on the path. Zurich had a lot of coffeehouses, and he spent afternoons in them, sipping the iced coffees, reading the newspapers, killing time with his friends. In quiet moments afterward, though, Einstein thought about physics and energy and other topics, and began getting hints of what might be wrong with the views he was being taught. All the types of energy that the Victorians had seen and shown to be interlinked—the chemicals and fires and electric sparks and blasting sticks—were just a tiny part of what might be. The energy domain was perceived as very large in the nineteenth century, but in only a few years Einstein would locate a source of energy that would dwarf what even the best, the most widely hunting of those Victorian scientists had found.

He would find a hiding place for further vast energy, where no one had thought to look. The old equations would no longer have to balance. The amount of energy God had set for our universe would no longer remain fixed. There could be more.

$$E=mc^2$$

Most of the main typographical symbols we use were in place by the end of the Middle Ages. Bibles of the fourteenth century often had text that looked much like telegrams:

> IN THE BEGINNING GOD CREATED THE HEAVEN
> AND THE EARTH AND THE EARTH WAS WITHOUT
> FORM AND VOID AND DARKNESS WAS UPON THE
> FACE OF THE DEEP

One change that took place at various times was to drop most of the letters to lowercase:

> In the beginning God created the heaven and the
> earth and the earth was without form and void and
> darkness was upon the face of the deep

Another shift was to insert tiny round circles to mark the major breathing pauses:

> In the beginning God created the heaven and the
> earth. And the earth was without form and void and
> darkness was upon the face of the deep.

Smaller curves were used as well, for the minor
breathing pauses:

> In the beginning, God created the heaven and the
> earth

Major symbols were locked in rather quickly once
printing began at the end of the 1400s. Texts began to be
filled in with the old ? symbols and the newer ! marks. It
was a bit like the Windows standard in personal computers driving out other operating systems.

Minor symbols took longer. By now we take them so
much for granted that, for example, we almost always
blink when we see the period at the end of a sentence.
(Watch someone when they're reading and you'll see it.)
Yet this is an entirely learned response.

For more than a thousand years, one of the world's
major population centers used this symbol Λ for addition, since it showed someone walking toward you (and
so was to be "added" to you), and Λ for subtraction.
These Egyptian symbols could easily have spread to become universally accepted, just as other Middle Eastern
symbols had done. Phoenician symbols, for example,
were the source of the Hebrew $\aleph$ and $\beth$—*aleph* and *beth*—
and also the Greek α and β—*alpha* and *beta*—as in our
word *alphabet*.

Through the mid-1500s there was still space for entre-
preneurs to set their own mark by establishing the remaining minor symbols. In 1543, Robert Recorde, an
eager textbook writer in England, tried to promote the
new-style "+" sign, which had achieved some popularity
on the Continent. The book he wrote didn't make his for-

tune, so in the next decade he tried again, this time with a symbol, which probably had roots in old logic texts, that he was sure would take off. In the best style of advertising hype everywhere, he even tried to give it a unique selling point: "... And to avoide the tediouse repetition of these woordes: is equalle to: I will sette ... a pair of parallels, or ... lines of one lengthe, thus: ═══════════ bicause noe .2. thynges, can be moare equalle. ..."

It doesn't seem that Recorde gained from his innovation, for it remained in bitter competition with the equally plausible // and even with the bizarre [; symbol, which the powerful German printing houses were trying to promote. The full range of possibilities proffered at one place or another include, if we imagine them put in the equation:

$e \parallel mc^2$

$e \longrightarrow mc^2$

$e \quad .æqus. \quad mc^2$

$e \,] \, [\, mc^2$

There was even my favorite:

$e \, ═══════════ \, mc^2$

Not until Shakespeare's time, a generation later, was Recorde's victory finally certain. Pedants and schoolmasters since then have often used the equals sign just to summarize what's already known, but a few thinkers had a better idea. If I say that 15+20=35, this is not very interesting. But imagine if I say:

(go **15** degrees west)

+

(then go **20** degrees south)

=

(you'll find trade winds that can fling you across the Atlantic to a new continent in **35** days).

Then I am telling you something new. A good equation is not simply a formula for computation. Nor is it a balance scale confirming that two items you suspected were nearly equal really are the same. Instead, scientists started using the = symbol as something of a telescope for new ideas—a device for directing attention to fresh, unsuspected realms. Equations simply happen to be written in symbols instead of words.

This is how Einstein used the "=" in his 1905 equation as well. The Victorians had thought they'd found all possible sources of energy there were: chemical energy, heat energy, magnetic energy, and the rest. But by 1905 Einstein could say, No, there is another place you can look where you'll find more. His equation was like a telescope to lead there, but the hiding place wasn't far away in outer space. It was down here—it had been right in front of his professors all along.

He found this vast energy source in the one place where no one had thought of looking. It was hidden away in solid matter itself.

$$E=mc^2$$

For a long time the concept of "mass" had been like the concept of energy before Faraday and the other nine-teenth-century scientists did their work. There were a lot of different material substances around—ice and rock and rusted metal—but it was not clear how they related to each other, if they did at all.

What helped researchers believe that there had to be some grand links was that in the 1600s, Isaac Newton had shown that all the planets and moons and comets we see could be described as being cranked along inside an immense, God-created machine. The only problem was that this majestic vision seemed far away from the nitty-gritty of dusty, solid substances down here on earth.

To find out if Newton's vision really did apply on Earth—to find out, that is, if the separate types of substance around us really were interconnected in detail—it would take a person with a great sense of finicky precision; someone willing to spend time measuring even tiny shifts in weight or size. This person would also have to be romantic enough to be motivated by Newton's grand vision—for otherwise, why bother to hunt for these dimly suspected links between all matter?

This odd mix—an accountant with a soul that could soar—might have been a character portrait of Antoine-Laurent Lavoisier. He, as much as anyone else, was the man who first showed that all the seemingly diverse bits of tree and rock and iron on earth—all the "mass" there is—really were parts of a single connected whole.

Lavoisier had demonstrated his romanticism in 1771 by rescuing the innocent thirteen-year-old daughter of his friend Jacques Paulze from a forced marriage to an un-couth, gloomy—yet immensely rich—ogre of a man. The reason he knew Paulze well enough to do the good deed for the daughter, Marie Anne, was that Paulze was his boss. The way he rescued Marie Anne was to marry her himself.

It turned out to be a good marriage, despite the difference in age, and despite the fact that soon after the handsome twenty-eight-year-old Lavoisier rescued Marie Anne, he shifted back to being immersed in the stupendously boring accountancy work he did for Paulze, within the organization called the "General Farm."

This was not a real farm, but rather an organization with a near monopoly on collecting taxes for Louis XVI's government. Anything extra, the Farm's owners could keep for themselves. It was exceptionally lucrative, but also exceptionally corrupt, and for years had at-tracted old men wealthy enough to buy their way in, but unable to do any detailed accounting or administration. It was Lavoisier's job to keep this vast tax-churning de-vice in operation.

He did that, head down, working long hours, six days a week on average for the next twenty years. Only in his spare time—an hour or two in the morning, and then one full day each week—did he focus on his science. But he called that single day his "jour de bonheur"—his "day of happiness."

Perhaps not everyone would comprehend why this was such a "bonheur." The experiments often resembled Lavoisier's ordinary accounting, only dragged out even longer. Yet the moment came when Antoine, in that irrational exuberance young lovers are known for, said his bride could now help him with a truly major experiment. He was going to watch a piece of metal slowly burn, or maybe just rust. He wanted to find out whether it would weigh more or less than it did before.

(Before going on, the reader might wish to actually guess: Let a piece of metal rust—think of an old fender or underbody panel on your car—and it ends up weighing

    a) less
    b) the same
    c) more

than it did before. Remember your answer.)

Most people, even today, probably would say it would weigh less. But Lavoisier, ever the cool accountant, took nothing on trust. He built an entirely closed apparatus, and he set it up in a special drawing room of his house. His young wife helped him: she was better at mechanical drawing than he was, and a lot better at English. (This would later be useful in keeping up with what the competition across the Channel was doing.)

They put various substances in their drawing room apparatus, sealed it tight, and applied heat or started an actual burn to speed up the rusting. Once everything had cooled down, they took out the mangled or rusty or otherwise burned-up metal and weighed it, and also carefully measured how much air may have been lost.

Each time they got the same result. What they found, in modern terms, was that a rusted sample does not weigh less. It doesn't even weigh the same. It weighs *more*.

This was unexpected. The additional weight was not from dust or metal shards left around in the weighing apparatus—he and his wife had been very careful. Rather, air has parts: there are different gases within the vapor we breathe. Some of the gases must have flown down and stuck to the metal. That was the extra weight he had found.

What was really happening? There was the same amount of stuff overall, yet now the oxygen that had been in the gases floating above was no longer in the air. *But it had not disappeared.* It had simply stuck on to the metal. Measure the air, and you would see it had lost some weight. Measure the chunk of metal, and you would see it had been enhanced—by exactly that same amount of weight the air had lost.

With his fussily meticulous weighing machine, Lavoisier had shown that matter can move around from one form to another, yet it will not burst in and out of existence. This was one of the prime discoveries of the 1700s—on a level with Faraday's realizations about energy in the basement of the Royal Institution a half century later. Here too, it was as if God had created a universe, and then said, I am going to put a fixed amount of mass in my domain, I will let stars grow and explode, I will let mountains form and collide and be weathered away by wind and ice; I will let metals rust and crumble. Yet throughout this the total amount of mass in my universe will never alter; not even to the millionth of an ounce; not even if you wait for all eternity. If a city were to be weighed, and then broken by siege, and its buildings burned by fire—if all the smoke and ash and broken ramparts and bricks were collected and weighed, there would be no change in the original weight. Nothing would have truly vanished, not even the weight of the smallest speck of dust.

To say that all physical objects have a property called their "mass," which affects how they move, is impressive, yet Newton had done it in the late 1600s. But to get enough detail to show exactly how their parts can combine or separate? That is the further step Lavoisier had now achieved.

Whenever France's scientists make discoveries at this level, they're brought close to the government. It happened with Lavoisier. Could this oxygen he'd helped clarify be used to produce a better blast furnace? Lavoisier had been a member of the Academy of Sciences and now was given funds to help find out. Could the hydrogen he was teasing out from the air with his careful measurements be useful in supplying a flotilla of balloons, capable of competing with Britain for supremacy in the air? He got grants and contracts for that as well.

In any other period this would have guaranteed the Lavoisiers an easy life. But all these grants and honors and awards were coming from the king, Louis XVI, and in a few years Louis would be murdered, along with his wife and many of his ministers and wealthy supporters.

Lavoisier might have avoided being caught up with the other victims. The Revolution was only at its most lethal phases for a few months, and many of Louis' closest supporters simply lived out those periods in quiet. But Lavoisier could never drop the attitude of careful measuring. It was part of his personality as an accountant; it was the essence of his discoveries in science.

Now it would kill him.

The first mistake seemed innocuous enough. Outsiders constantly bothered members of the Academy of Sciences, and long before the Revolution, one of them, a Swiss-born doctor, had insisted that only the renowned Lavoisier would be wise enough, and understanding enough, to judge his new invention. The device was something of an early infrared scope, allowing the doctor to detect the shimmering heat waves rising from the top of a candle, or of a cannonball, or even—on one proud occasion, when he'd lured the American representative to his chambers—from the top of Benjamin Franklin's bald head. But Lavoisier and the Academy turned him down. From what Lavoisier had heard, the heat patterns that the doctor was searching for couldn't be measured with precision, and to Lavoisier that was anathema. But the Swiss-born hopeful—Dr. Jean-Paul Marat—never forgot.

The next mistake was even more closely linked to Lavoisier's obsession with measurement. Louis XVI was helping America fund its revolutionary war against the British, an alliance that Benjamin Franklin had been central in sustaining. There were no bond markets, so to get the money Louis had to turn to the General Farm. But taxes already were high. Where could they go to get more?

In every period of incompetent administration France has suffered—and Louis's successors in the 1930s would have given him a good run—there almost always has been a small group of technocrats who've decided that since no one who was officially in power was going to take charge, then they would have to do it themselves. Lavoisier had an idea. Think of the measuring apparatus in his drawing room, the one where he and Marie Anne had been able to keep exact track of everything going in and out. Why not enlarge it, wider and wider, so that it encompassed all of Paris? If you could track the city's incomings and outgoings, he realized, you could tax them.

There once had been a physical wall around Paris, but it dated from medieval times, and had long since become nearly useless for taxation. Tollgates were crumbling, and many areas were so broken that smugglers could just walk in.

Lavoisier decided to build another wall, a massive one, where everyone could be stopped, searched, and forced to pay tax. It cost the equivalent, in today's currency, of several hundred million dollars; it was the Berlin Wall of its time. It was six feet high, of heavy masonry, with dozens of solid tollgates and patrol roads for armed guards.

Parisians hated it, and when the Revolution began, it was the first large structure they attacked, two days before the storming of the Bastille: they tore at it with firebrands and axes and bare hands till it was almost entirely gone. The culprit was known, as an antiroyalist broadsheet declared: "Everybody confirms that M. Lavoisier, of the Academy of Sciences, is the 'beneficent patriot' to whom we owe the . . . invention of imprisoning the French capital. . . ."

Even this he might have survived. A mob's passions are brief, and Lavoisier hurriedly tried to show he was on their side. He personally directed the gunpowder mills that supplied the Revolutionary Armies; he tried to have the Academy of Sciences show new, reformist credentials by getting rid of the grand tapestries in its Louvre offices. He even seemed to be succeeding—until one never-forgiving figure from his past emerged.

By 1793, Jean-Paul Marat was head of a leading faction in the National Assembly. He'd suffered years of poverty because of Lavoisier's rejection: his skin was withered from an untreated disease, his chin unshaven, his hair neglected. Lavoisier by contrast was still hand-some: his skin was smooth; his build was strong.

Marat didn't kill him immediately. Instead, he made sure Paris's citizens were constantly reminded of the wall, this living, large-scale summary of everything Marat hated about the class-smug Academy. He was a magnificent speaker—along with Danton and, in recent history, Pierre Mendès-France, among the finest France has produced. ("I am the anger, the just anger, of the people and that is why they listen to me and believe in me.") The only sign of Marat's tension—barely visible to listeners watching his confident posture, right hand on his hip, left arm casually extended on the desk in front of him—was a slight nervous tapping of one foot on the ground. When Marat denounced Lavoisier, he embod-ied the very principle that Lavoisier had demonstrated. For was it not true that everything balances? If you seem to destroy something in one place, it's not really de-stroyed. It just appears somewhere else.

In November 1793 Lavoisier got word he was going to be arrested. He tried hiding in the abandoned parts of the Louvre, roaming through the Academy's empty offices there, but after four days he gave up, and walked—with Marie Anne's father—to the Port Libre prison.

If he looked out his window of the Port Libre ("Our address is: first floor hall, number 23, room at the end"), he could see the great classical dome of the Observatory, a landmark over one century old, and now closed by Revolutionary orders. At least at night, when the guards ordered candles blown out in Lavoisier's prison, the stars were visible above its dome.

There were transfers to other prisons; the trial itself was on May 8. A few prisoners tried to speak, but the judges laughed at them. Marat's bust was on a shelf looking down on the accused. That afternoon, twenty-eight of the onetime millionaires from the General Farm were taken to what's now the Place de la Concorde. Their hands were tied behind their backs. It was a steep climb up to the working level of Dr. Guillotin's instrument. Most seem to have been quiet, though one of the older men "was led to the scaffold in a pitiful state." Paulze was third; Lavoisier was fourth. There was about a minute after each beheading: not to clean the blade, but to clear away the headless bodies.

With Lavoisier's work, the conservation of mass was on its way to being established. He had played a central role in helping to show that there was a vast, interconnected world of physical objects around us. The substances that fill our universe can be burned, squeezed, shredded, or hammered to bits, but they won't disappear. The different sorts floating around just combine or recombine. The total amount of mass, however, remains the same. It would be the perfect match to what Faraday later found: that energy is conserved as well. With all of Lavoisier's accurate weighing and chemical analysis, researchers were able to start tracing how that conservation happened in practice—as with his working out how oxygen molecules cascaded from the air to stick to iron. Breathing was more of the same, simply a means of shifting oxygen from the outer atmosphere to the inside of our bodies.

By the mid-1800s, scientists accepted the vision of energy and mass as being like two separate domed cities. One was composed of fire and crackling battery wires and flashes of light—this was the realm of energy. The other was composed of trees and rocks and people and planets—the realm of mass.

Each one was a wondrous, magically balanced world;

35

each was guaranteed in some unfathomable way to keep its total quantity unchanged, even though the forms in which it appeared could vary tremendously. If you tried to get rid of something within one of the realms, then something else within that same realm would always pop up to take its place.

Everyone thought that nothing connected the two realms, however. There were no tunnels or gaps to get between the blocking domes. This is what Einstein was taught in the 1890s: that energy and mass were different topics; that they had nothing to do with each other.

Einstein later proved his teachers wrong, but not in the way one might expect. It is common to think of science as building up gradually from what came before. The telegraph is tinkered with and turns into the telephone; a propeller airplane is developed, and studied, and then improved planes are built. But this incremental approach does not work with deep problems. Einstein did find that there was a link between the two domains, but he didn't do it by looking at experiments with weighing mass and seeing if somehow a little bit was not accounted for, and might have slipped over to become energy. Instead he took what seems to be an immensely roundabout path. He seemed to abandon mass and energy entirely, and began to focus on what appeared to be an unrelated topic.

He began to look at the speed of light.

c Is for celeritas

$$E=mc^2$$

"c" is different from what we've looked at so far. "E" is the vast domain of energies, and "m" is the material stuff of the universe. But "c" is simply the speed of light.

It has this unsuspected letter for its name probably out of homage for the period before the mid 1600s when science was centered in Italy, and Latin was the language of choice. *Celeritas* is the Latin word meaning "swiftness" (and the root of our word *celerity*).

What this chapter looks at is how "c" came to play such an important role in $E=mc^2$: how this particular speed—what might seem an arbitrary number—can actually control the link between all the mass and all the energy in the universe.

For a long time even measuring the speed of light was considered impossible. Almost everyone was convinced that light traveled infinitely fast. But if that were so, it could never have been used in a practical equation. Before anything more could be done—before Einstein could have possibly thought of using "c"—someone had to confirm that light travels at a finite speed, but that wouldn't be easy.

37

. . .

Galileo was the first person to clearly conceive of measuring the speed of light, well before he was undergoing house arrest in his old age, nearly blind. By the time he published, though, he was too old to carry out the experiment himself, and the Inquisition had given strict orders controlling where he went. That was little more than a challenge to him and his friends. A few years after his death, when members of an academy for experimental studies in Florence came to hear of his work, they let it be known that they would do the observations he had proposed.

The idea was as simple as all Galileo's work had been. Two volunteers were to stand holding lanterns on hillsides a mile apart one summer evening. They would open their lantern shutters one after the other, and then time how much of a delay there was for the light to cross the valley.

The experiment was a good idea, but the technology of the time was too poor to get any clear result. Galileo had been aware of the need in other experiments to breathe regularly, so as not to speed his heartbeat when an experiment was under way, for he used his pulse to measure short intervals of time. But that evening, probably in the hills outside of Florence, the volunteers found the light was too quick. All they noticed was a quick blur, a movement that seemed instantaneous. This could have been seen as a failure, and for most people it was just another proof that light traveled at infinite speed. But the Florentines didn't accept that it meant Galileo's speculations were wrong. Rather, the Academy concluded, it would just have to be left to someone from a future generation to find a way to time this impossibly fast burst.

In 1670, several decades after Galileo's death in 1642, Jean-Dominique Cassini arrived in Paris to take up his position as head of the newly established Paris Observatory. There was a lot of new construction to supervise, and he could sometimes be seen in the street doing that—not far from the shadows of the Port Libre prison, where Lavoisier in the next century would await his death—but his most important task was to shake some life into French science. He also had a personal incentive to make the new institution succeed, for his name wasn't actually Jean-Dominique but Giovanni Domenico. And he wasn't French, but newly arrived from Italy, and although the king was on his side, and the funding was said to be guaranteed, who knew how long that really would last?

Cassini sent emissaries to the fabled observatory of Uraniborg, on an island in the Danish straits not far from Elsinor Castle. Their goal was to fix the coordinates of Uraniborg, which would help in measuring distances for navigation; they might also find skilled researchers to recruit from other observatories. The founder of the Uraniborg observatory, Tycho Brahe, had made the observations on which Kepler and even Newton based their work. Brahe had created unimagined luxuries: there were exotic species of trees, gardens with artificial canals and fish ponds around the central castle, an impressive intercomlike communication system, and rotating automata that terrified local peasants; there were even rumors of an automatic flush toilet.

Cassini's right-hand man, Jean Picard, reached Uraniborg in 1671, sailing the misty waters from Copenhagen. He was excited about finally getting to see the fabled stronghold—then dismayed at finding it was a complete wreck.

Those sophisticated findings that had impressed

Kepler dated from almost a century before. The observatory's founder had been a powerful personality, but when he had died, no one really took over. Everything was decayed or broken when Picard arrived: the fish ponds filled in, the quadrants and celestial globe long gone; only a few of the foundation stones of the main house were still recognizable.

Picard did get his readings, however, and also managed to bring back to Paris a bright twenty-one-year-old Dane named Ole Roemer. Others might be humbled to meet the great Cassini when they arrived back, for Cassini was a world authority on the planet Jupiter, and especially on the orbits of its satellites as they rotated around the planet. But although today we think of Denmark as a small nation, at that time it ran an empire that encompassed a good stretch of northern Europe. Roemer was cockily proud, enough to try making his own name.

It's doubtful whether Cassini was especially pleased with the upstart. It had taken a long time for him to make the switch from Giovanni Domenico to Jean-Dominique. He had accumulated numerous detailed observations of Jupiter's satellites, and he was certainly going to use them to maintain his worldwide reputation. But what if Roemer plundered his findings to prove that the conclusions Cassini was drawing from them were all wrong?

The reason this was possible was that there was a problem with the innermost moon of Jupiter, the one called Io. It was supposed to travel around its planet every $42\frac{1}{2}$ hours. But it never stuck honestly to schedule. Sometimes it was a little quicker, sometimes a little slower. There was no discernible pattern anyone could tell.

But why? To solve the problem, Cassini insisted on more measurements and calculations. The effort might be exhausting for the observatory's director, and of course it might entail more staff and more equipment and more funds and more patronage, and all that embarrassing public exposure, but if that was necessary, it could be done. To Roemer, however, what was needed was not the sort of complex measurements only middle-aged administrators could manage. What was needed was the brilliance, the inspiration, that a young outsider applying his mind could provide.

And this is what Roemer did. Everyone—even Cassini—assumed that the problem was in how Io traveled. Possibly it was ungainly and wobbled during its orbit; or possibly there were clouds or other disturbances around Jupiter that obscured Io unevenly. Roemer reversed the problem. Cassini had made observations of Io, and the observations showed that something about its orbit was not smooth. But why should the flaw be assumed to rest far away near Jupiter? The question wasn't how Io was traveling, thought Roemer.

It was how Earth was traveling.

To Cassini, this couldn't possibly matter. Although he may once have considered a different possibility, like almost everyone else, he was convinced that light traveled as an instantaneous flash. Any fool could see that. Hadn't Galileo's own experiment shown that there was no evidence to the contrary?

Roemer ignored all that. Suppose—just suppose—that light did take some time to travel the great distance from Jupiter. What would that mean? Roemer imagined he was straddling the solar system, watching the first flicker of Io's light burst out from behind the planet Jupiter, and rush all the way to Earth. In the summer, for example, if Earth was closer to Jupiter, the light's journey would be shorter, and Io's image would arrive sooner. In the winter of the same year, though, Earth could have swung around to the other side of the solar system. It would take a lot longer for Io's signal to reach us.

Roemer went through Cassini's stacked years of observations, and by the late summer of 1676 he had his solution: not just a hunch, but an exact figure for how many extra minutes light took to fly that extra distance when Earth was far from Jupiter.

What should he do with such a finding? By protocol, Roemer should have let Cassini present it as his own work, and simply nod modestly, perhaps, when the observatory chief paused to remark that he couldn't have done it without the help of this young man whose future career was worth watching.

Roemer didn't go along with that. In August, at the esteemed public forum of a journal all serious astronomers read, he proclaimed a challenge. Astronomy is an exact science, and even seventeenth-century tools were good enough to determine that the satellite Io was scheduled to appear from behind Jupiter on the coming November 9, sometime in the late afternoon. By Cassini's reasoning, it would be detected at 5:27 P.M. on that day. That was what he extrapolated from when it had last been clearly sighted, in August.

Roemer declared that Cassini was going to be wrong. In August, he explained, the earth had been close to Jupiter. In November it would be farther away. There would be nothing visible at 5:27—the light, though fast, would still be on its way, since it had to travel that extra distance. Even at 5:30 or 5:35 it wouldn't have made it across the solar system. Only at 5:37 precisely would anyone be able to get their first sighting on November 9.

There are many ways to make astronomers happy. A new supernova is good; a renewed grant from the government is good; tenure is extremely good. But an out-and-out battle between two of your distinguished colleagues? It was heaven. Roemer had thrown down his challenge partly out of pride, but also because he knew that Cassini was a much better political operator than he was. Roemer would only be able to claim credit if his prediction was so clear that Cassini and his minions couldn't wangle out of it if they were wrong.

The prediction was announced in August. On November 9, observatories in France and across Europe had their telescopes ready. 5:27 P.M. arrived. No Io.

5:30 arrived. Still no Io.

5:35 P.M.

And then it appeared, at 5:37 and 49 seconds exactly.

And Cassini declared he had not been proven wrong! (Spin-doctoring was not invented in the era of television.) Cassini had lots of supporters, and support him they did. Who'd ever said Io was expected at 5:25? That had only been Roemer, they now declared. Besides, everyone realized Io's arrival time was never certain. It was so far away, so hard to see exactly, that perhaps those clouds from Jupiter's upper atmosphere were producing a distorting haze; or maybe the high angle of its orbit was what made definite observations so difficult. Who knew?

In the usual history of science accounts, it's not supposed to happen this way. Roemer had performed an impeccable experiment, with a clear prediction, yet Europe's astronomers still did not accept that light traveled at a finite speed. Cassini's supporters had won: the official line remained that the speed of light was just a mystical, unmeasurable figure; that it should have no impact on astronomical measurements.

Roemer gave up, and went back to Denmark, spending many years as the director of the port of Copenhagen. Only fifty years later—after a further generation had passed, and Jean-Dominique Cassini was gone—did further experiments convince astronomers that Roemer had been right. The value he had estimated for light's speed was close to the best current estimate, which is about 670,000,000 mph. (In fact the exact speed is a fraction higher, but for convenience we'll round it off to 670 million for the rest of this book.)

To emphasize how big a number this is, at 670,000,000 mph, you could get from London to Los Angeles in under 1/20th of a second. That explains why Galileo's experiment could not detect the time it took light to cross a valley outside Florence; the distance was much too small.

There's another comparison: Mach 1 is the speed of sound, about 700 mph. A 747 jet travels at a little under Mach 1. The space shuttle, after full thrust, can surpass Mach 20. The asteroid or comet that splashed a hole in the ocean floor and destroyed the dinosaurs impacted at about Mach 70.

The number for "c" is Mach 900,000.

This vast speed leads to many curious effects. Let someone irritatingly speak into a cell phone just a few tables away from you in a restaurant, and it seems as if you're hearing his voice almost as soon as the words leave his mouth. But air only can carry a sound wave at the lowly speed of Mach 1, whereas radio signals shooting upward from the cell phone travel as fast as light. The person on the receiving end of the phone—even if she's hundreds of miles away—will hear the words *before* they've trundled the few yards through the air in the restaurant to reach you.

To see why Einstein chose this particular value for his equation, we need to look closer, at light's inner properties. The story leaves the epoch of Cassini and Roemer far behind, and picks up in the late 1850s, the period just before the American Civil War, when the now elderly Michael Faraday began to correspond with James Clerk Maxwell, a slender young Scot still in his twenties.

It was a difficult time for Faraday. His memory was failing and he could often barely get through a morning without extensive notes to remind him of what he was supposed to do. Even worse, Faraday also knew that the world's great physicists, almost all of whom had gone to elite universities, still patronized him. They accepted his practical lab findings, but nothing else. To standard physicists, when electricity flowed through a wire it was basically like water flowing through a pipe: once the underlying mathematics was finally worked out, they believed, it would not be too different from what Newton and his numerous mathematically astute successors could describe.

Faraday, however, still went on about those strange circles and other wending lines from his religious upbringing. The area around an electromagnetic event, Faraday held, was filled with a mysterious "field," and stresses within that field produced what were interpreted as electric currents and the like. He insisted that sometimes you could almost see their essence, as in the curving patterns that iron filings take when they are sprinkled around a magnet. Yet almost no one believed him—except, now, for this young Scot named Maxwell.

At first glance the two men seemed very different. In his many years of research, Faraday had accumulated over 3,000 paragraphs of dated notebook entries on his experiments, from investigations that began early every morning. Maxwell, however, quite lacked any ability to get a timely start to the day. (When he was told that there was mandatory 6 A.M. chapel at Cambridge University, the story goes that he took a deep breath, and

James Clerk Maxwell

said, "Aye, I suppose I can stay up that late.") Maxwell also had probably the finest mathematical mind of any nineteenth-century theoretical physicist, while Faraday had problems with any conventional math much beyond simple addition or subtraction.

But on a deeper level the contact was close. Although Maxwell had grown up in a great baronial estate in rural Scotland, the family name had until recently simply been Clerk, and it was only from an inheritance on the maternal side that they'd acquired the more distinguished Maxwell to tack on. When young James was sent away to boarding school in Edinburgh, the other children—stronger in build, cockily confident with their big-city ways—picked on him: week after week, year after year. James never expressed any anger about it; just once, he quietly remarked: "They never understood me, but I understood them." Faraday also still carried the wounds from his experiences with Sir Humphry Davy in the 1820s, and would relapse into a quiet, watching solitude almost instantly after he'd finished an evening as an apparently ebullient speaker at one of the Royal Institution public lectures.

When the young Scot and the elderly Londoner corresponded, and then later when they met, they cautiously made contact of a sort they could share with almost no one else. For beyond the personality similarities, Maxwell was such a great mathematician that he was able to see beyond the surface simplicity of Faraday's sketches. It was not the childishness that less gifted researchers mocked it for. ("As I proceeded with the study of Faraday, I perceived that his method . . . was also a mathematical one, though not exhibited in the conventional form of mathematical symbols.") Maxwell took those crude drawings of invisible force lines seriously. They were both deeply religious men; they both appreciated this possibility of God's immanence in the world.

Back in his 1821 breakthrough, and then in much research after, Faraday had shown ways in which electricity can be turned into magnetism, and vice versa. In the late 1850s, Maxwell extended that idea, into the first full explanation of what Galileo and Roemer had never understood.

What was happening inside a light beam, Maxwell began to see, was just another variation of this back-and-forth movement. When a light beam starts going forward, one can think of a little bit of electricity being produced, and then as the electricity moves forward it powers up a little bit of magnetism, and as that magnetism moves on, it powers up yet another surge of electricity, and so on like a braided whip snapping forward. The electricity and magnetism keep on leapfrogging over each other in tiny, fast jumps—a "mutual embrace," in Maxwell's words. The light Roemer had seen hurtling across the solar system, and which Maxwell saw bouncing off the stone towers at Cambridge, was merely a sequence of these quick, leapfrogging jumps.

It was one of the high points of nineteenth-century science; Maxwell's equations summarizing this insight became known as one of the greatest theoretical achievements of all time. But Maxwell was always slightly dissatisfied with what he'd produced. For how exactly did this strangely leapfrogging light wave braid itself along? He didn't know. Faraday didn't know. No one could explain it for sure.

Einstein's genius was to look closer at what these skittering light waves meant, even though he had to do it largely on his own. He had the confidence to do this: his final high school preparation in Aarau really had been superb, and he'd grown up in a family that encouraged him to always question authority. By the 1890s, when Einstein was a student, Maxwell's formulations were usually taught as a received truth. But Einstein's main professor at the Zurich polytechnic, unimpressed with theory in physics, refused even to teach Maxwell to his undergraduates. (It was Einstein's resentment at being treated like this that led him to address that professor mockingly as *Herr Weber* rather than the expected *Herr Professor Weber*—a slight that Weber avenged by refusing to write a proper letter of recommendation for Einstein, leading to his years of isolation at the patent office job.)

When Einstein cut classes to go to the coffeehouses in Zurich, it was often with accounts of Maxwell's work in hand. He began to explore the leapfrogging of light waves that Maxwell had first uncovered. If light was a wave like any other, Einstein mused, then if you ran after it, could you catch up?

An example from surfing will show the problem. When you're first out in the water, trying not to let everyone on shore see how scared you are, the waves slosh past you. But once you force yourself to stand up on your surfboard, you can glide shoreward as the wave of water pushing you seems to be standing still around you. Be bold enough—or foolhardy enough—to do this in the extreme surf off Hawaii, and an entire curving tube of water might seem to be at rest around, above, and beside you.

Only in 1905 did the full insight hit Einstein. Light waves are different from everything else. A surfer's water wave can appear to hold still, because all the parts of the wave take up a steady position in relation to one another. That's why you can glance out from your surfboard, and see a hovering sheet of water. Light is not like that, however. Light waves keep themselves going only by virtue of one part moving forward and so powering up the next part. (The electricity part of the light wave shimmers forward, and that "squeezes" out a magnetic part; then that magnetic part, as it powers up, creates a further "surge" of electricity so the rushing cycle starts repeating.) Whenever you think you're racing forward fast enough to have pulled up next to a light beam, look harder and you'll see that whatever part you thought you were close to is powering up a further part of the light beam that is still hurtling away from you.

To catch up with a streak of light and see it standing still would be like saying, "I want to see the blurred arcs of a thrilling juggling act, but only if the balls are not moving." You can't do it. The *only* way you'll see a blur from the juggling balls is if they're moving fast.

Einstein concluded that light can exist only when a light wave is actively moving forward. It was an insight that had been lurking in Maxwell's work for over forty years, but no one had recognized it.

This new realization about light changed everything; for the speed of light becomes the fundamental speed limit in our universe: nothing can go faster.

It's easy to misunderstand this. If you were traveling at 669,999,999 mph, couldn't you pump in more fuel, and go the few mph faster—to 670,000,000, and then to 670,000,0001—to take you past the speed of light? But the answer is that you can't, and it's not a quirk about the present state of earthly technology.

A good way to recognize this is to remember that light isn't just a number, it is a physical process. There's a big difference. If I say that –273 (negative 273) is the lowest number that there is, you could rightly answer that I'm wrong: that –274 is lower, and –275 is lower yet, and that you can keep on going forever. But suppose we were dealing with temperatures. The temperature of a substance is a readout of how much its inner parts are moving, and at some point they're going to stop vibrating entirely. That happens at about –273 degrees on the centigrade scale, and that's why –273 degrees is said to be "absolute zero" when you're talking about temperature. Pure numbers might be able to go lower, but physical things can't: a coin or a snowmobile or a mountain can't vibrate any less than not vibrating at all.

So it is with light. The 670,000,000 mph figure that Roemer measured for the light speeding down from Jupiter is a statement about what that light is like. It's a physical "thing." Light will always be a quick leapfrogging of electricity out from magnetism, and then of magnetism leaping out from electricity, all swiftly shooting away from anything trying to catch up to it. That's why its speed can be an upper limit.

. . .

It's an interesting enough observation, but a cynic might say, so what if there *is* an ultimate speed limit? How should that affect all the solid objects that move around within the universe? You can clamp a label saying "Warning: No Speed Over 670,000,000 mph Can Be Achieved," on the signs by a busy road, but the traffic whirring past will be unaffected.

Or will it? This is where Einstein's whole argument finally circles back: where he showed that light's curious properties—the fact that it inherently squiggles away from you, and is therefore the ultimate speed limit—really enters into the nature of energy and mass. An example modified from one that Einstein himself used can suggest how it might happen.

Suppose a super space shuttle is blasting along very close to the speed of light. Under normal circumstances, when that space shuttle is going slowly, the fuel energy that's pumped into the engines would just raise its speed. But things are different when the shuttle is right at the very edge of the speed of light. It can't go much faster.

The pilot of the shuttle doesn't want to accept this, and starts frantically leaping up and down on the thruster control to get the vessel to go faster. But of course the pilot sees any light beam that's ahead still squirting out of reach at the full speed of "c." That's what light does for any observer. Despite the pilot's best efforts, the shuttle is not gaining on it. So what happens?

Think of frat boys jammed into a phone booth, their faces squashed hard against the glass walls. Think of a parade balloon, with an air hose pumping into it that can't be turned off. The whole balloon starts swelling, far beyond any size for which it was intended. The same thing would happen to the shuttle. The engines are roaring with energy, but that can't raise the shuttle's

ANCESTORS OF E = MC²

speed, for nothing goes faster than light. But the energy can't just disappear, either.

As a result, the energy being pumped in gets "squeezed" into becoming mass. Viewed from outside, the solid mass of the shuttle starts to grow. There's only a bit of swelling at first, but as you keep on pouring in energy, the mass will keep on increasing. The shuttle will keep on swelling.

It sounds preposterous, but there's evidence to prove it. If you start to speed up small protons, which have one "unit" of mass when they're standing still, at first they simply move faster and faster, as you'd expect. But then, when they get close to the speed of light, an observer really will see the protons begin to change. It's a regular event at the accelerators outside of Chicago, and at CERN (the European center for nuclear research) near Geneva, and everywhere else physicists work. The protons first "swell" to become two units of mass—twice as much as they were at the start—then three units, then on and on, as the power continues to be pumped in. At speeds of 99.9997 percent of "c," the protons end up 430 times bigger than their original size. (So much power is drained from nearby electricity stations that the main experiments are often scheduled to run late at night, so that nearby residents won't complain about their lights dimming.)

What's happening is that energy that's pumped into the protons or into our imagined shuttle has to turn into extra mass. Just as the equation states: that "E" can become "m," and "m" can become "E."

That's what explains the "c" in the equation. In this example, as you push up against the speed of light, that's where the linkages between energy and mass become especially clear. The figure "c" is merely a conversion factor telling you how that linkage operates.

Whenever you link two systems that developed separately, there will need to be some conversion factor. To go from centigrade to Fahrenheit, you multiply the centigrade figure by 9/5, and then add 32. To go from centimeters to inches there's another rule: you multiply the centimeters by 0.3937.

The conversion factors seem arbitrary, but that's because they link measurement systems that evolved separately. Inches, for example, began in medieval England, and were based on the size of the human thumb. Thumbs are excellent portable measuring tools, since even the poorest individuals could count on regularly carrying them along to market. Centimeters, however, were popularized centuries later, during the French Revolution, and are defined as one billionth of the distance from the equator to the North Pole, passing by Paris. It's no wonder the two systems don't fit together smoothly.

For centuries, energy and mass had also seemed to be entirely separate things. They evolved without contact. Energy was thought of as horsepower or kilowatt hours; mass was measured in pounds or kilos or tons. No one thought of connecting the units. No one glimpsed what Einstein did, that there could be a "natural" transfer between energy and mass, as we saw with the shuttle example, and that "c" is the conversion factor linking the two.

The reader might wonder when we'll get to the theory of relativity. The answer is that we've already been using it! All these points about a speeding shuttle and its expanding mass are central to what Einstein published in 1905.

Einstein's work changed the two separate visions scientists had taken from the nineteenth-century work on conservation laws. Energy isn't conserved, and mass isn't conserved—but that doesn't mean there is chaos.

Instead, there's actually a deeper unity, for there's a link between what happens in the energy domain and what happens in the seemingly distinct mass domain. The amount of mass that's gained is always going to be balanced by an equivalent amount of energy that's lost.

Lavoisier and Faraday had seen only part of the truth. Energy does not stand alone, and neither does mass. But the sum of mass plus energy will always remain constant.

This, finally, is the ultimate extension of the separate conservation laws the eighteenth- and nineteenth-century scientists had once thought complete. The reason this effect had remained hidden, unsuspected, all the time before Einstein, is that the speed of light is so much higher than the ordinary motions we're used to. The effect is weak at walking speed, or even at the speed of locomotives or jets, but it's still there. And as we'll see, the linkage is omnipresent in our ordinary world: all the energy is held quiveringly ready inside even the most ordinary substances.

Linking energy and mass via the speed of light was a tremendous insight, but there's still one more detail to get clear. A famous cartoon shows Einstein at a board, trying out one possibility after another: $E=mc^1$, $E=mc^2$, $E=mc^3$, ... But he didn't really do it that way, arriving at the squaring of "c" by mere chance.

So why did the conversion factor turn out to be c^2?

E=mc²

Enlarging a number by "squaring" it is an ancient procedure. A garden that has four paving slabs on one edge, and four on the other, doesn't have eight stones in it. It has 16.

The convenient shorthand that summarizes this action of building up a "square"—of multiplying a number by itself—went through almost the same range of permutations as did the Western typography for the equals sign. But why should it appear in physics equations? The story of how an equation with a "squared" in it came to be plucked from all other possibilities for representing the energy of a moving object takes us back to France once more—to the early 1700s—and the generation halfway between Roemer and Lavoisier.

In February 1726, the thirty-one-year-old playwright François-Marie Arouet was convinced he'd successfully gate-crashed the establishment in France. He'd risen from the provinces to receive grants from the king, acceptance at the homes of noblemen, and one evening was even being dined at the gated home of the Duc de

Sully. A servant interrupted the meal: there was a gentleman outside to see Arouet.

He went out and probably had a moment to recognize the carriage of the Chevalier de Rohan, an unpleasant, yet staggeringly rich man whom he'd mocked in public when they'd recently attended a play at the Comédie Française. Then de Rohan's bodyguards got to work, beating Arouet while de Rohan watched, delighted, from inside his carriage, "supervising the workers," as he later described it. Somehow, Arouet managed to get back inside the gates, and into de Sully's home. But instead of sympathy or even outrage, Arouet encountered only laughter. De Sully and his friends were amused: a preposterous wordsmith had been put in his place by someone who really mattered. Arouet vowed to avenge himself; he would challenge de Rohan to a duel, and kill him.

That was getting too serious. De Rohan's family had a word with the authorities; there was a police hunt; Arouet was arrested, then put in the Bastille.

When he finally got out he crossed the Channel, falling in love with England, and especially—estate agents take note—with the bucolic wonderland of Wandsworth, far from the grime of the busy city. He was exhilarated to find that there was a new concept in the air, the works of Newton, which represented what could be the opposite of the ancient, locked-in aristocratic system he'd known in France.

Newton had created a system of laws that seemed to detail, with superb accuracy, how every part of our universe moved about. The planets swung through space at a rate and in directions that Newton's laws described; a cannonball fired in the air would land exactly where Newton's calculations of its trajectory showed that it would land.

It really was as if we were living inside a vast windup clock, and all the laws Newton had seen were simply the gears and cogs that made it work. But if we could demand a rational explanation of the grand universe beyond our planet, Arouet wondered, shouldn't we be able to demand the same down here on earth? France had a king who demanded obedience, on the grounds that he was God's regent on earth. Aristocrats got authority from the king, and it was impious to question this. But what if the same analysis used in science by Newton could be used to reveal the role of money or vanity or other hidden forces in the political world as well?

By the time Arouet went back to Paris, three years later, he had begun pushing his new ideas, in private letters and printed essays. In a world of clear, levelheaded analysis of true forces, his humiliation outside the gates of de Sully would never have been allowed. Arouet would support Newton's new vision accordingly his whole life long. He was a good supporter to have, for Arouet was only the name he'd been born with. He'd already largely put it to the side for the pen name by which he was better known: Voltaire.

But even a skilled writer, however eager to push a particular thinker, can't shift a nation on his own. Voltaire needed to be able to place his talents within a switching center that could multiply them. The king's Academy of Sciences was too backward-looking; too locked into the old guard's way of thinking. The salons of Paris wouldn't do either. The usual hostesses were rich enough to keep a tame poet or two ("If you neglect to enroll yourself among the courtesans," Voltaire observed, "you are . . . crushed"), but there was no space for a real thinker. He needed help. And he found it.

. . .

He'd actually met her without realizing it, fifteen years before, visiting her father when she was just a girl. Emilie de Breteuil's family lived overlooking the Tuileries gardens in Paris, in an apartment with thirty rooms and seventeen servants. But although her brothers and sisters turned out as expected, Emilie was different, as her father wrote: "My youngest flaunts her mind, and frightens away the suitors. . . . We don't know what to do with her."

When she was sixteen they brought her to Versailles, but still she stood out. Imagine the actress Geena Davis, Mensa member and onetime action-film star, trapped in the early eighteenth century. Emilie had long black hair and a look of perpetual startled innocence, and although most other debutante types wanted nothing more than to use their looks to get a husband, Emilie was reading Descartes's analytic geometry, and wanted potential suitors to keep their distance.

She'd been a tomboy as a child, loving to climb trees, and she was also taller than average, and—best of all— since her parents had been worried she'd end up clumsy, they'd paid for fencing lessons for years. She challenged Jacques de Brun, whose position was roughly equivalent to head of the king's bodyguard detail, to a demonstration duel, in public, on the fine wood floor of the great Hall of Arms. She was fast enough, and strong enough, with the thrusts and parries, to remind any overeager suitors that they would be wise to leave her alone.

Her intellect left her isolated at Versailles, for there was no one with whom she could share her excitement about the wondrous insights she was discovering through the work of Descartes and other researchers. (At least there were certain advantages in being immersed in equations—she found it easy to memorize cards at the blackjack table.)

When Emilie was nineteen, she chose one of the least objectionable courtiers as a husband. He was a wealthy soldier named du Châtelet, who would conveniently be on distant campaigns much of the time. It was a pro forma arrangement, and in the habit of the time, her husband accepted her having affairs while he was away. There were a number of lovers, one of the closest being a onetime guards officer, Pierre-Louis Maupertuis, who had resigned his post, and was in the process of becoming a top physicist. Their courtship had begun in studying calculus and more advanced work together, but he was leaving on a polar expedition, and in 1730s France, no twenty-something young woman—however bright, however athletic—would be allowed to go along.

Now Emilie was at loose ends. Where could she turn for warmth? She had a few desultory affairs while Maupertuis was ordering his final supplies, but who, in France, could fill Maupertuis's place? Enter Voltaire.

"I was tired of the lazy, quarrelsome life led at Paris," Voltaire recounted later, ". . . of the privilege of the king, of the parties and cabals among the learned. . . . In the year 1733 I met a young lady who happened to think nearly as I did. . . ."

She met Voltaire at the opera, and although there was some overlap with Maupertuis, that was no problem. Voltaire composed a stirring poem for Maupertuis, complimenting him as a modern-day argonaut, for his boldness in venturing to the far north for science; he then wrote a romantic poem to du Châtelet, comparing her to a star, and noting that he, at least, was not so faithless as to exchange her for some expedition to the Arctic pole. It wasn't entirely fair to Maupertuis, but du Châtelet didn't mind. Anyway, what could Voltaire do? He was in love.

Emilie du Châtelet
PAINTING BY MAURICE QUENTIN DE LA TOUR. LAUROS-GIRAUDON

And so, finally, was she. This time she wasn't going to let it go. She and Voltaire shared deep interests: in political reform, in the fun of fast conversation ("she speaks with great rapidity," one of her earlier lovers had written, ". . . her words are like an angel"); above all, they shared a drive to advance science as much as they could. Her husband had a château, at Cirey, in northeastern France. It had been in the family since before Columbus went to America, and now was largely shuttered up; abandoned. Why not use that as a base for genuine scientific research in France? They got to work, and Voltaire soon wrote to a friend that Mme. du Châtelet

> . . . is changing staircases into chimneys and chimneys into staircases. Where I ordered the workmen to construct a library, she tells them to place a salon. . . . She's planting lime trees where I'd planned to place elms, and where I only planted herbs and vegetables . . . only a flowerbed will make her happy.

Within two years it was complete. There was a library comparable to that of the Academy of Sciences in Paris, the latest laboratory equipment from London, and there were guest wings, and the equivalent of seminar areas, and soon there were visits from the top researchers in Europe. Du Châtelet had her own professional lab, but the wall decorations in her reading areas were original paintings by Watteau; there was a private wing for Voltaire, yet also a discreet passageway conveniently connecting his bedroom with hers. (Arriving one time when she didn't expect him, he discovered her with another lover, and she tried putting him at ease by explaining that she'd only done this because she knew he hadn't been feeling well and she hadn't wanted to trouble him while he needed his rest.)

The occasional visitors from Versailles who came to scoff saw a beautiful woman willingly staying inside, working at her desk well into the evening, twenty candles around her stacks of calculations and translations; advanced scientific equipment stacked in the great hall. Voltaire would come in, not merely wanting to gossip about the court—though, being Voltaire, he was unable to resist this entirely—but also to compare Newton's Latin texts with some of the latest Dutch commentaries.

At several times she came close to jump-starting future discoveries. She performed a version of Lavoisier's rust experiment, and if the scales she'd been able to get machined had been only a bit more accurate, she might have been the one to come up with the law of the conservation of mass, even before Lavoisier was born.

The Cirey team kept up a supporting correspondence with other new-style researchers; supplying them with whatever evidence, diagrams, calculations might be needed. The scientific visitors such as Koenig and

Bernoulli sometimes stayed for weeks or months at a time. Voltaire was pleased that crisp, Newtonian science was gaining ground through their efforts. But when he and du Châtelet engaged in their teasing, their mock battling, it wasn't the case of a worldly, widely read man deciding when to let his young lover win. Du Châtelet was the real investigator of the physical world, and the one who decided that there was one key question that had to be turned to now: What is energy?

She knew that most people felt energy was already sufficiently well understood. Voltaire had covered the seemingly ordained truths in his own popularizations of Newton: the central factor to look for when you're analyzing how objects make contact is simply the product of their mass times their velocity, or their mv^1. If a 5-pound ball is going 10 mph, it has 50 units of energy.

But du Châtelet knew that there had once been a famous competing view to Newton's, due to Gottfried Leibniz, the great German diplomat and natural philosopher. For Leibniz, the important factor to focus on was mv^2. If a 5-pound ball is going at 10 mph, it has 5 times 10^2, or 500 units of energy.

Which view was true? It might seem a mere quarreling over definitions, but there was something deeper going on behind it. We're used to science being separated from religion, but in the seventeenth and eighteenth centuries it wasn't.

Newton felt that highlighting where mv^1 occurs would prove that God had to exist. If two identical beer wagons crash head-on, there's an almighty bang, and possibly some grinding as their bumpers crumple into each other, but then there's stillness. Right before they hit there was a lot of mv^1 in the universe: the two speeding carts were each loaded with the stuff. One cart had been going full speed due east, for example; the other had been going full speed due west. After they hit, though, and had become stationary chunks of wood and metal, the two separate parts of the v^1 were gone. The "going due east" had exactly canceled out the "going due west."

In Newton's view, this meant that all the energy the carts had once possessed had now vanished. A hole had been created, leading out from our visible universe. Since collisions like this happen all the time, if we live within a great, coglike clockwork, that clock would always need winding. But look around you. We don't find that as the years pass, fewer and fewer objects are able to move. That's the proof. The fact that the universe continues operating was, in Newton's view, a sign that God's reassuring hand was reaching in, to nurture us and to support us; to supply all the motive forces we otherwise lost.

For Voltaire that was enough. Newton had spoken, and who was he to argue with Newton, and anyway it seemed such a magnificent vision—and it was backed by such distressingly complicated geometry and calculus—that it was wisest just to nod in confirmation and accept it. But du Châtelet spent a long time in her room with the Watteau paintings, and then at the candle-edged writing table, working through Leibniz's contrary arguments for herself.

Along with various abstract geometric arguments, Leibniz had also focused on the way that Newton's approach left gaps in the world. Diplomats can be sarcastic. He wrote: "According to [Newton's] doctrine, God Almighty wants to wind up his watch from time to time: otherwise it would cease to move. He had not, it seems, sufficient foresight to make it a perpetual motion."

It turned out that concentrating on measurements of energy as being mv² avoided this problem. The mv² of a cart going due west might be, say, 100 units of energy, and the mv² of a second cart going on a collision course due east might be another 100 units. For Newton the two hits canceled each other out, but for Leibniz they added up. When the two carts hit, all the energy they carried remained busily in existence, sending metal parts bouncing and rebounding, heating up the wagon wheels, generally creating an ongoing, reverberating jangle.

In this view of Leibniz's, nothing is lost. The world runs itself; there are no holes or sluicegates where causality and energy rushes away, so that only God would be able to pour them back in. We're alone. God might have been needed at the very beginning, but no longer.

Du Châtelet found some attraction in this analysis, but also recognized why it had languished in the decades since Leibniz had proposed it. This view was too vague; matching Leibniz's personal biases, but without enough objective proof. It was also, as Voltaire got great satisfaction showing in his novel *Candide*, a strangely passive view; suggesting that no fundamental improvements to our worldly condition could be made.

Du Châtelet was known for being burstingly quick in conversation, but at Versailles that had been because she was surrounded by fools, while at Cirey that was the only way to get a word in with Voltaire. When it came to her original work, she was much more methodical, taking her time. After going through the first arguments by Leibniz, and then the standard critiques against them, she—and various specialists she brought in to help—didn't leave it there, but started looking wider, for some practical evidence that would help her make a choice. To Voltaire she was clearly "wasting" her time, but for du Châtelet it was one of the peak moments of her life: the research machine she had established at Cirey was finally being used to its full capacity.

She and her colleagues found the decisive evidence in the recent experiments of Willem 'sGravesande, a Dutch researcher who'd been letting weights plummet onto a soft clay floor. If the simple $E=mv^1$ was true, then a weight going twice as fast as an earlier one would sink in twice as deeply. One going three times as fast would sink three times as deep. But that's not what 'sGravesande found. If a small brass sphere was sent down twice as fast as before, it pushed *four* times as far into the clay. If it was flung down three times as fast, it sank *nine* times as far into the clay.

Which is just what thinking of $E=mv^2$ would predict. Two squared is four. Three squared is nine. The equation's operation really did seem, in some strange way, fundamental to nature.

'sGravesande had a solid result but wasn't enough of a theoretician to put it all together. Leibniz was a top theoretician but had lacked this detailed experimental finding—his opting for mv^2 had been a bit of a guess. Du Châtelet's work on this topic bridged the gap. She deepened Leibniz's theory, and then embedded the Dutch results within it. Now, finally, there was a strong justification for viewing mv^2 as a fruitful definition of energy.

Her publications had a great effect. Du Châtelet had always been a clear writer, and it helped that Cirey was looked up to as one of the few truly independent research centers. Most English-speaking scientists automatically took Newton's side, while German-speaking ones tended to be just as dogmatically for Leibniz. France had always been the crucial swing vote in the middle, and Du Châtelet's voice was key in finally tilting the debate.

After publishing her work she paused—to take care of her family's finances and to consider what research topic to do next. There were travels with Voltaire, and she was amused that the new generation of courtiers at Versailles had no idea that she was one of the leading interpreters of modern physics in Europe, or that in her spare time she had published original translations of Aristotle and Virgil. Occasionally it would slip, when she did a burst of probability calculations for the gaming table.

Time passed, and they went back to Cirey. The lime trees were growing ("in this, our delightful retreat," as she wrote), and she had even let Voltaire have his vegetable garden. And then, as she hurriedly wrote in a letter to a friend

3 April, 1749

Château de Cirey

I am pregnant and you can imagine . . . how much I fear for my health, even for my life . . . giving birth at the age of forty.

It was the one thing she couldn't control. She'd had children shortly after her marriage, but she had been twenty years younger, and even then it had been dangerous. Being this much older, survival was not very likely. Doctors of the time had no awareness that they should wash their hands or instruments. There were no antibiotics to control the inevitable infection; nothing like oxytocin, which can control uterine bleeding. She didn't rage at the clear incompetence of her era's doctors; she just said to Voltaire that it was sad leaving before she was ready. The length of time before her was very clear: the labor was expected in September. She'd always worked long hours; now she sped up, the candles at the desk where she wrote sometimes burning till dawn.

On September 1, 1749, she wrote to the director of the king's library, stating that he would find in the accompanying package the now complete draft of a major commentary she was doing on Newton. Three days later, the birth began; she survived that, but infection set in, and within a week she died.

Voltaire was beside himself: "I have lost the half of myself—a soul for which mine was made."

In time the focus on energy as being proportional to mv^2 began to seem second nature to physicists. Voltaire's polemical skills, passing on the legacy of his lover, helped give it an even stronger boost. In the following century, Faraday and others used mv^2—this quantity that might transform but never totally disappeared—as they built up their visions of the conservation of all energy. Du Châtelet's analysis and writing had been an indispensable step, though in time her role came to be forgotten; partly because each new generation of scientists tends to be generally neglectful of their past; partly, perhaps, because it was unsettling to hear that a woman could have directed such a large research effort and helped shape the course of subsequent thought.

The big question, though, is *why*. Why is squaring the velocity of what you measure such an accurate way to describe what happens in nature?

One reason is that the very geometry of our world

often produces squared numbers. When you move twice as close toward a reading lamp, the light on the page you're reading doesn't simply get twice as strong. Just as with the 'sGravesande experiment, the light's intensity increases *four* times.

When you are at the outer distance, the light from the lamp is spread over a larger area. When you go closer, that same amount of light gets concentrated on a much smaller area.

The interesting thing is that almost *anything* that steadily accumulates will turn out to grow in terms of simple squared numbers. If you accelerate on a road from 20 mph to 80 mph, your speed has gone up by four times. But it won't take merely four times as long to stop if you apply brakes and they lock. Your accumulated energy will have gone up by the square of four, which is sixteen times. That's how much longer your skid will be.

Imagine that skid hooked up to some sort of energy collector. A car that's racing along at four times another one's speed, really will generate—really does carry along—sixteen times as much energy. If someone tried to measure energy as simply equal to mv^1, they'd miss all this. Only by concentrating on mv^2 do these important aspects come out.

Over time, physicists became used to multiplying an object's mass by the square of its velocity (mv^2) to come up with such a useful indicator of its energy. If the velocity of a ball or rock was 100 mph, then they knew that the energy it carried would be proportional to its mass times 100 squared. If the velocity is raised as high as it could go, to 670 million mph, it's almost as if the ultimate energy an object will contain should be revealed when you look at its mass times c squared, or its mc^2. This isn't a proof, of course, but it seemed so natural, so "fitting," that when the expression mc² did suddenly appear within Einstein's more detailed calculations, it helped make more plausible his startling conclusion that the seemingly separate domains of energy and mass could be connected, and that the symbol "c"—the speed of light—was the bridge. (For the reader interested in Einstein's actual derivations, the Web site for this book, davidbodanis.com, goes through some of his reasoning.)

The c² is crucial in saying how this link operates. If our universe were created differently—if c² were a low value—then a small amount of mass would only be transformed into an equally small puff of energy. But in our real universe, and viewed from the small, ponderously rotating planet to which we're consigned, c² is a huge number. In units of mph, c is 670 million, and so c² is 448,900,000,000,000,000. Visualize the equals sign in the equation as a tunnel or bridge. A very little mass gets enormously magnified whenever it travels through the equation and emerges on the side of energy.

This means that mass is simply the ultimate type of condensed or concentrated energy. Energy is the reverse: it is what billows out as an alternate form of mass under the right circumstances. As an analogy, think of the way that a few wooden twigs going up in flames can produce a great volume of billowing smoke. To someone who'd never seen fire, it would be startling that all that smoke was "waiting" inside the wood. The equation shows that any form of mass can, in theory, be manipulated to expand outward in an analogous way. It also says this will happen far more powerfully than what you would get by simple chemical burning—there is a much greater "expansion." That enormous conversion factor of 448,900,000,000,000,000 is how much any mass gets magnified, if it's ever fully sent across the "=" of the equation.

69

PART 3

The Early
Years

$$E=mc^2$$

When Einstein published $E=mc^2$ in 1905, the equation was at first almost entirely ignored. It simply did not fit in with what most other scientists were doing. The great insights from Faraday and Lavoisier and all the rest were available, but no one else was putting them together this way—hardly anyone even had a hint that one could try.

The world's dominant industries were steel and railways and dyes and agriculture, and that's what ordinary researchers concentrated on. A few universities had specialized labs for more theoretical work, but much of that was in areas that wouldn't have been too surprising to Newton over two centuries before: there were treatises on conventional optics, and sound, and elasticity. There was a little fresh work, on the new and puzzling radio waves, and in areas related to radioactivity, but Einstein was mostly on his own.

We can date to within a month or so the moment when he first saw that E would equal mc^2. Einstein finished writing his initial paper on relativity by late June 1905, and had the addendum with the equation ready for printing in September, so he probably first realized it

73

some time in July or August. It would likely have been either on one of his walks, or at home after his day job at the patent office. Often his infant son, Hans Albert, was around when he worked, but that wouldn't have been a problem. Visitors recount Einstein contentedly working in the living room of his small apartment, while rocking his one-year-old's bassinet with his free hand, humming or singing to him as needed.

What guided Einstein was that, in his mid-twenties, he found the unknown intriguing. He felt compelled to comprehend what might have been intended for our universe by The Old One (as he referred to his notion of God).

"We are in the position," Einstein explained later, "of a little child entering a huge library, whose walls are covered to the ceiling with books in many different languages. The child knows that someone must have written those books. It does not know who or how. It does not understand the languages in which they are written. The child notes a definite plan in the arrangement of the books, a mysterious order, which it does not comprehend but only dimly suspects."

When the chance came to reach through the gloom, and pluck out The Old One's book that had the shimmering equation $E=mc^2$ written on its pages, Einstein had been willing to take it.

The reasoning Einstein followed to come up with his extraordinary observation—that mass and energy are one—had begun with the seemingly irrelevant observation that no one could ever catch up with light. But that led, as our space shuttle example suggested, to the insight that energy pouring into a moving object could end up making an outside observer see its mass swell.

The argument could also apply in reverse: under the right circumstances an object should be able to pour out energy, generating it from its own mass.

Starting in the 1890s, a few years before Einstein wrote out his equation, a number of investigators had actually seen hints of how this might occur. Several metal-streaked ores that had been brought back from the Congo and Czechoslovakia and other places were found, in laboratories in Paris and Montreal and elsewhere, to be spraying out some sort of mysterious energy beams. If the pebbles were used up as they did this, it wouldn't have been too surprising—one could think that the process was some sort of ordinary burning. But by the best measurements of the time, the energy beams seemed to be pouring out without the pebbles changing in any way.

Marie Curie was one of their first investigators, and indeed in 1898 coined the word *radioactivity* for this active spurting out of radiation. Yet even she, at first, had no understanding that these metals achieved their power by sucking immeasurably tiny portions of their mass out of existence, and switching that mass into the greatly magnified form of sprayed energy. The amounts seemed beyond credibility: a palm-sized chunk of these ores could spray out many trillions of high-speed alpha particles every second, and repeat this for hours and weeks and months, without any loss of weight that anyone could measure.

Later, after Einstein was famous, he met Curie several times, but he never understood her—after one hiking trip he described her as being cold as a herring and constantly complaining. In fact, she had a passionate nature and was deeply in love with an elegant French scientist who was married to someone else. The reason she complained on the hiking trip may have been because that she was slowly dying of cancer. Radium was one of these scarcely understood new metals, and Curie had been working with it for years.

The minute traces of radium powder, which she had carried unknowingly on her blouse and hands as she walked across the muddy cobblestones of 1890s Paris and later, had been pouring out energy in accord with the then-unsuspected equation, barely shrinking at all, for thousands of years. They had been spray-releasing part of themselves without getting used up back when they were deep underground in the Belgian mines in the Congo; they continued through her years of experiments, ultimately giving her this killing cancer. More than seventy years later, the dust would still be alive and could squirt out poisonous radiation onto any archivists who were examining her office ledger, or even the cookbooks at her home.

The amount of dust Curie had scattered was measured in millionths of an ounce. But that had been enough, in accord with Einstein's equation, for the radioactive dust to slam into the DNA in her bones, producing the leukemia of which she died; to slam upward, only a fraction more feebly, into the detecting Geiger counters of any such startled archivists so many decades later.

Einstein's equation showed how large the result could be. To work it out for any chunk of mass, take the great speed of light and square that to get an even more immense number. Then, multiply that by the amount of mass you're looking at, and that's how much energy, exactly, the mass will be able to pour out.

It's easy to miss how powerful that idea is. For $E=mc^2$ says nothing about what sort of mass can fit into the equation! Under the proper circumstances, *any* substance can have its mass exploded outward as energy.

This is the power that's around us, encased within the most ordinary rocks and plants and streams. A single page of this book, weighing only a few grams, seems to be just an innocuous, stable mix of cellulose fibers and ink. But if that ink and cellulose could ever be shifted into the form of pure energy, there would be a roaring eruption, greater than that of a large power station exploding. It's easier to access that power in uranium than in ordinary paper—as we'll see later—but that's simply a limitation of our current technology.

The greater the mass being transformed, the more fearsome the power released. Put a single pound of mass into the "m" slot, and after multiplying by the vast 448,900,000,000,000,000 value of c^2, the equation promises that, in principle, you could get over 10 billion kilowatt hours of energy. This is comparable to a huge power station. That's how a small atomic bomb—with a core small enough to fit in your cupped hands—could heave out enough energy to rip open streets and buried fuel lines; to shatter street after street of brick buildings; to tear open the bodies of tens of thousands of soldiers and children and teachers and bus drivers.

A uranium bomb works when less than 1 percent of the mass inside it gets turned into energy. An even larger amount of matter, compressed into a floating star, can warm a planet for billions of years, just by seemingly squeezing part of itself out of existence, and turning those fragments of once-substantial matter into glowing energy.

In 1905, when Einstein first wrote out his equation, he was so isolated that he prepared the main relativity article without footnotes. That's almost unheard of in science. The one acknowledgment Einstein did put in was to his loyal friend Michele Besso, a thirty-something mechanical engineer, working at the patent office, who happened to be the author's friend. Even in 1905 physicists complained of being overburdened. Einstein's articles appeared in a distinguished journal—he'd been keen enough on his career to stay connected by submitting review articles—but one after another, the physicists turning through the journal either skimmed or just ignored this exceptional misfit of an article.

At one point Einstein tried applying for a junior teaching position at the university in Bern, as a way out of the patent office. He sent off the relativity article he was so proud of, along with others he'd written. He was rejected. A little later he applied to a high school, again offering his services as a teacher. The equation was sealed in the envelope with the rest of his application forms. There were twenty-one applicants, and three got called in for interviews. Einstein wasn't one of them.

In time a few scientists did begin to hear of his work, and then jealousy set in. Henri Poincaré was one of the glories of Third Republic France, and, along with David Hilbert in Germany, one of the greatest mathematicians in the world. As a young man Poincaré had written up the first ideas behind what later became chaos theory; as a student, the story goes, he'd once seen an elderly woman on a street corner knitting, and then, thinking about the geometry of her knitting needles as he walked along the street, he'd hurried back and told her that there was another way she could have done it: he'd independently come up with purling.

By now, though, he was in his fifties, and although he could still get some fresh ideas, he increasingly didn't have the energy to develop them. Or maybe it was more than that. Middle-aged scientists often say that the problem isn't a lack of memory, or the ability to think quickly. It's more a fearfulness at stepping into the unknown. For Poincaré had once had the chance of coming close to what Einstein was doing.

In 1904 he'd been in the large group of disoriented European intellectuals invited to the World's Fair being held in St. Louis. (Max Weber, the German sociologist, was also there, and he was so startled by the raw energy he saw in America—he described Chicago as being "like a man whose skin has been peeled off"—that it helped jolt him out of a depression he'd been suffering for years.) At the fair, Poincaré had actually given a lecture on what he'd labeled a "theory of relativity," but that name is misleading for it only skirted around the edges of what Einstein would soon achieve. Possibly if Poincaré had been younger he could have pushed it through to come up with the full results Einstein reached the next year, including the striking equation. But after that lecture, and then the exhausting schedule his St. Louis hosts had for him, the elderly mathematician let it slide. The fact that so many French scientists had turned away from Lavoisier's hands-on approach and instead insisted on a sterile overabstraction only made it harder for Poincaré to be immersed in practical physics.

By 1906, realizing that this young man in Switzerland had opened up an immense field, Poincaré reacted with the coldest of sulks. Instead of looking closer at this equation, which he could have considered a stepchild, and bringing it in to his Paris colleagues for further development, he kept a severe distance; never speaking of it; seldom mentioning Einstein's name.

Other contemporaries did examine Einstein's work more closely, but tended to miss, at first, such key points as why Einstein selected "c" as being so central. They could understand if relativity and the equation had come from some fresh experimental results; if Einstein had built some new-style apparatus in a laboratory to look more closely at what Marie Curie or others were finding, and so had discoveries which no one else did. But what they could not grasp was that he didn't *have* any labs. The "latest findings" he worked with came from scientists who'd died decades or even centuries before. But that didn't matter. Einstein hadn't come up with his ideas by patiently putting together a range of new results. Instead, as we saw, he just spent a long time "dreamily" thinking about light and speed and what was logically possible in our universe and what wasn't. But it only seemed "dreamy" to outsiders who didn't understand him. What he ended up accomplishing was one of the major intellectual achievements of all time.

For centuries after the birth of mathematically guided science around the seventeenth century, humans thought that they had the main lines of the universe described, and that although there were further details to work out, the "commonsense" properties of the world around us could be taken for granted. We lived in a world where objects kept a constant mass as they moved around; where time advanced smoothly, and everyone could agree about where we were in its flow.

Einstein saw that the universe was different from what everyone had thought. It was, he realized, as if God had restricted us to a small playpen—the surface of the Earth—and had even let us think that what we observed from it was all that really occurred. Yet all the while, stretching further out—around us all the time if we were able to see it—was a further domain, where our intuition no longer applied. Only pure thought would allow us to see what happened there.

The fact of energy and mass being interchangeable, as shown in $E=mc^2$, is only one of these fuller consequences. There are others as well, and to recognize them, it helps to imagine a world where instead of the uppermost speed limit being the speed of light at 670 million, it is instead an easy 30 mph. What does Einstein's 1905 relativity paper say we'll see?

The first striking thing we would see if we entered that world follows from the space shuttle example. Cars would have their ordinary weight when they were waiting patiently at a red light, but once the light changed to green, they would bulk up in mass as they got faster. It would happen to pedestrians and joggers and bicyclists and indeed everything that moved. A schoolchild, who might weigh 100 pounds on her bicycle when waiting at a corner, would bulk up to 230 pounds once she had pedaled up to 27 mph. If she was fast, or had a downhill slope to help and got up to 29.97 mph, she'd soon have a mass of over 2,000 pounds. Her bicycle would swell up just as much. As soon as she stopped pedaling both she and her bicycle would immediately come down to their original, static weight.

At the same time, cars, bicycles, and even pedestrians would undergo another change. Depending on the position from which it was being watched a 12-foot car would undergo distortions such as parts of it appearing smaller (and its position shifted) as it roared toward us. At 29.9 mph parts of it would be tiny. The driver and passengers inside would appear to have shrunk just as much, and, again, as soon as they stopped, would settle back and return to their normal appearance.

As the cars hurried by, we'd not only see them as getting more massive and changing size, we'd also notice that time seemed to be slowing down inside of them. If the driver reached to turn on the CD player, we'd see his hand move in extreme slow motion. Once the player was on and the sound was coming out, we'd hear the sound waves transmitted with painful slowness, transforming even early Michael Jackson warbles into heavy, dirgelike chants.

In this view of the universe there is no "true" perspective—some sort of traffic helicopter hovering above this odd city—from which one can assert that, yes, the cars are undergoing these strange changes, but the bystanders who aren't moving are unchanged and clearly "normal." For why should the bystanders have some favored status, while it's only the moving cars that are changing? In fact, the drivers of the cars, or the schoolgirl on her bicycle, will have no sensation that they're changing. The bicyclist will look around her, and see that her handlebars and her body and her backpack haven't become heavier. Rather, to her it's the people left behind who will seem weird. *They*'ll be the ones whose mass has swelled.

The passengers in the car will agree. Their CD player is fine, they'll say, and the young Michael Jackson is warbling along as quickly as ever. It's the people outside the car who seem slow, with hotel doormen seeming to lift their arms in laborious heaviness, then puffing out their cheeks like stately deep-sea fish whenever they blow a whistle to hail a taxi.

These effects are summarized in relativity by saying that when someone watches an object recede away from them, that object will be seen to undergo mass dilation, length changes, and time dilation. The bystanders will see it in the car; the driver of the car, looking back, will see it in the bystanders.

The first time one reads of this, it seems like nonsense. Even Einstein found it hard to accept—as with the inexplicable tension he felt in his long talk with Michele Besso, on the summer day when he was still trying to work out these relations. But it's only hard to accept because we never actually interact with each other at speeds close to the 670 million mph of light (and the effects are too slight to notice at our ordinary speeds). Think, for example, of a portable music player at a picnic. To someone standing next to it, it's loud. To someone who walks a few hundred yards away, the music is soft. We accept that there's no answer about how loud it "really" is. But that's simply because we're capable of walking quickly enough to cover that few hundred yards in a brief time. To an ant or some smaller creature, one that took many generations to migrate far enough from the music player to detect any change in its volume, our view—that music can appear to be at different volumes to different observers—would seem crazy.

The Web site gives the details of how physicists show that all this must follow from such simple observations as light's constancy. But there are a number of ordinary objects around us, which always do work at the high speeds where these effects become apparent. The electrons that shoot from the back of traditional TV sets to the screen at the front, for example, travel so fast that they really will respond to us as if they've grown in mass as they travel. Engineers have to take that into account when they design the magnets that focus the screen's image. If they didn't we would see a blur.

The Global Positioning System (GPS) of navigational satellites, which fly overhead and beam down location signals for cars and jets and hikers, are also traveling so fast that from our perspective, time on board them seems to be slow. The circuits in the handheld GPS location devices we use to locate our positions, or in the larger GPS devices that banks use to synchronize payments, are programmed to correct for this—in exact accord with the equations Einstein worked out in 1905.

Einstein never especially liked the label *relativity* for what he'd created. He thought it gave the wrong impression, suggesting that anything goes: that no exact results any longer occur. That's not so. The predictions are precise.

The label is also misleading because all Einstein's equations are cohesive, and exactly linked up. Although each of us might view things in the universe differently, there will be enough synchronization where these different views join to ensure that it all fits. The old notions that mass never changes and that time flows at the same rate for everyone made sense when people only noticed the ordinary, slow-moving objects around them. In the true wider universe, however, they're not correct—but there are exact laws to explain how they change.

This is an achievement that has occurred very few times in history. Imagine being able to make a shimmering crystalline model, small enough to hold in your closed fist. Now open your hand—and see the entire universe soar out; glowing into full existence. Newton was the first person to have done that, back in the 1600s: conceiving a complete system of the world, that could be described in but a handful of equations, yet also contained the rules for how to move out from the summary and go on to creating the full world.

Einstein was the next.

Just to make the bond more impressive, both Einstein and Newton achieved much of their work in impossibly brief periods in their mid-twenties. For Newton, back at his mother's Lincolnshire farm after his university had been closed because of the Plague, there were about eighteen months in which he did fundamental work on developing calculus, conceiving the law of universal gravitation, as well as working on key concepts for a mechanics that would apply throughout the universe. For Einstein, in a period of under eight months in 1905—and while still putting in full days at the patent office, Monday through Saturday—there was his first theory of relativity, and $E=mc^2$, as well as his work that helped lay the path for lasers, computer chips, key aspects of the modern pharmaceutical and bio-engineering industry, and all Internet switching devices. He really was, as Newton described himself when similarly in his mid-twenties, "in the prime of my age for invention." In each area, Einstein pushed beyond what was known; he unified fields that had remained separate, questioning assumptions that everyone until then had simply accepted.

The few researchers around 1905 who had uncovered a small part of what he later deduced had no chance of matching him. Poincaré got closer than almost anyone else, but when it came to breaking our usual assumptions about time's flow or the nature of simultaneity, he backed off, unable to consider the consequences of such a new view.

Why was Einstein so much more successful? It's tempting to say it was just a matter of being brighter than everyone else. But several of Einstein's Bern friends were highly intelligent, while someone like Poincaré would have been off the scale on any IQ test. Thorstein Veblen once wrote a curious little essay that I think gets at a deeper reason. Suppose, Veblen began, a young boy learns that everything in the Bible is true. He then goes to a secular high school, or university, and is told that's wrong. "What you learned at your mother's knee is entirely false. What we teach you here, however, will be entirely true." Some students would say, Oh, fine, I'll accept that. But others will be more suspicious. They'd been fooled once before, taking on faith an entire traditional world. They're not going to be fooled again. They would learn what was on offer, but always hold it critically, as just one possibility among others. Einstein was Jewish, and even though his immediate family wasn't observant, this meant he was immersed in a culture with different views about personal responsibility, justice, and belief in authority than the standard German and Swiss consensus.

There's more, though. When Einstein was a little boy, he was fascinated with how magnets worked. But instead of being teased about it by his parents, they accepted his interest. How *did* magnets work? There had to be a reason, and that reason had to be based on another reason, and maybe if you traced it all the way, you'd reach . . . what would you reach?

At one time, in the Einstein household, there had been a very clear answer to what would ultimately be reached. When his grandparents had been growing up, most Jews in Germany were still close to traditional Orthodoxy. It was a world suffused by the Bible, as well as by the crisply rational accumulated analysis of the Talmud. What counted was to push through to the very edge of what was knowable, and comprehend the deepest patterns God had decreed for our world. Einstein had gone through an intense religious period when he was approaching his teens, though by the time he was at the Aarau high school that literal belief was gone. Yet the desire to see the deepest underpinnings was still there, as was the trust that you would find something magnificent waiting if you made it that far. There was a waiting "slot": things could be clarified, and in a comprehensible, rational way. At one time the slot had been filled in by religion. It could easily enough be extended now to science. Einstein had great confidence that the answers were waiting to be found.

It also helped that Einstein had the space to explore his ideas. The patent job meant that he didn't have to churn out academic papers ("a temptation to superficiality," Einstein wrote, "which only strong characters can resist"), but rather he could work on his ideas for as long as it took. Most of all, his family trusted him, which is a great boost to confidence, and they also encouraged a playful, distancing tone. It's just what's needed for "stepping back" from ordinary assumptions, and imagining such oddities as a space shuttle pushed up against a barrier at the speed of light, or someone chasing toward a skedaddling beam of light.

His sister, Maja, later gave a hint of this gently self-teasing tone. When Einstein got in a temper as a little child, she recounted, he sometimes threw things at her. Once it was a large bowling ball; another time he used a child's hoe "to try to knock a hole" in her head. "This should suffice," she commented, "to show that it takes a sound skull to be the sister of an intellectual." When she described the high school Greek teacher who complained that nothing would ever become of her brother, she added: "And in fact Albert Einstein never did attain a professorship of Greek grammar."

To crank it all forward, there need to be driving tensions, and these Einstein had aplenty. There was the failure of being in his mid-twenties, isolated from other serious scientists, when university friends were already making careers for themselves. There was also thunderous guilt from seeing the difficulties his father was having in his own business career. Einstein had grown up with his father fairly prosperous in the electrical contracting business in Munich, but when Einstein was a teenager, possibly because key contracts stopped being given to Jewish firms, his father had moved the family to Italy to set up again. In the move, and in a series of near-successes that never quite made it, his father was exhausted in paying back loans to a brother-in-law, the constantly nagging Uncle Rudolf "The Rich" (as Einstein mockingly called him). It wrecked his father's health; yet through it all the family had insisted on finding the money to pay for Einstein to study. ("He is oppressed by the thought that he is a burden on us, people of modest means," as his father had remarked in the 1901 letter.) There was a huge obligation for Einstein to show he had been worth it after that.

Eventually a few other physicists did begin to pay attention to Einstein, sometimes visiting Bern to talk over the equation and other results. It was just what Einstein and Besso had hoped for, but it also meant that they started being pulled apart. For Einstein was gradually going beyond the ideas his best friend could follow. Although Besso was bright, he'd chosen a life in industry. ("I prodded him very much to become a [university teacher], but I doubt . . . he'll do it. He simply doesn't want to.") Besso couldn't follow the next level.

Besso adored his younger friend, and had gone out of his way to help him back when Einstein was still a student. He even tried, hard, in their evenings sharing Gruyère and sausages and tea, to keep up with the further ideas Einstein was seeing now. Einstein himself was kind about the growing distance from his friends. He never declared to Besso that he was no longer interested in him. They continued country walks, stops for a drink, musical evenings, and practical jokes with the others. But it's a bit like two old school friends breaking off once both have started moving separate ways at university, or in their first jobs afterward. Each one would really like things not to be like that, but everything they care about now is pulling them apart. They can talk about the old days when they're together, but the enthusiasm is forced, even though neither of them wants to admit it.

A similar distancing happened with Einstein's wife, Mileva. She'd been a physics student with him, and very bright. Men in the sciences rarely marry fellow specialists—how many are there?—and Einstein was almost smug to his college friends about how lucky he'd been. His first letters to her had started neutrally:

> Zurich, Wednesday [16 February 1898]
>
> I have to tell you what material we covered. . . .
> Hurwitz lectured on differential equations (exclusive of partial ones), also on Fourier series. . . .

But the relationship developed, as extracts from a series of letters written in August and September 1900 show:

> Once again a few lazy and dull days flitted past my sleepy eyes, you know, such days on which one gets up late because one cannot think of anything proper to do, then goes out until the room has been made up. . . . Then one hangs around and looks halfheartedly forward to the meal. . . .
>
> However things turn out, we are getting the most delightful life in the world. Beautiful work, and together. . . .
>
> Be cheerful, dear sweetheart. Kissing you tenderly, your
>
> Albert

The life they shared started out happily. His wife wasn't going to be at his level, but she really was a good student—on the university final exams where he scored 4.96, she came close, with a 4.0, and she certainly could have followed his work. (The myth that she had been responsible for his key work stems from nationalist Serb propaganda in the 1960s, as her family had originally been from near Belgrade.) But once their children came, and with their income so low that they only had part-time help, all the traditional sexism took over. When educated friends came to visit, his wife would try to join in, but this is never easy with an attention-frantic three-year-old son on your lap. You want to stay a part of the conversation, but after too many interruptions for getting toys and drawing pictures and picking up spilled food, the guests no longer stop their talk to recap things and bring you in. You're left out.

Einstein finally left the patent office—though even when he did, in 1909, his chief was mystified as to why this young man was willing to turn his back on such a good career. He was finally offered a position in the Swiss university system, and then after a stint in Prague—where he played music and engaged in discussions at a salon that occasionally included a shy young man named Franz Kafka—Einstein ended up as a professor in Berlin. His success had now isolated him almost completely from his Bern friends. He was legally separated from his wife, and only occasionally got to see his adored two children.

By that time, Einstein was taking his personal work in a different direction. The equation $E=mc^2$ was just a small part of the entire special theory of relativity. By 1915, he'd perfected an even grander theory, so powerful that the entire special theory was just one small part of that. (The Epilogue gives some highlights of that 1915

Mileva and Albert Einstein
MAX FLÜCKIGER, *EINSTEIN IN BERN*. COPYRIGHT BY PAUL HAUPT, BERN.
AIP EMILIO SEGRÈ VISUAL ARCHIVES

work—"Compared with this problem, the original theory of relativity is child's play.") He would be involved with the equation only once more, briefly, when he was a much older man.

At this point there's a major shift in our story. The equation's first theoretical development was over; Einstein's personal contribution fades away. Europe's physicists accepted that $E=mc^2$ was true: that, in principle, matter could be transformed so that the frozen energy it was composed of could be let out. But no one knew how actually to get that to happen.

There was one hint. It came in the strange objects that Marie Curie and others were investigating: the dense metals of radium, and uranium, and other substances, which were somehow able to pour out energy week after week, month after month; never using up whatever "hidden" source of supply they contained inside.

A number of laboratories began to study how that might be happening. But to see what mechanisms were creating these great outwellings of energy, it wouldn't be enough to continue looking at the surface of things, simply measuring the weight or color or surface chemical properties of the mysteriously warm radium or uranium.

Instead, the researchers would have to go within, deep into the very heart of these substances. That, ultimately, would show how the energy that $E=mc^2$ promised could be accessed. But what would they find, as they tried to peer into the smallest, inner structures within ordinary matter?

Into the Atom

8

$$E=mc^2$$

University students in 1900 were taught that ordinary matter—bricks and steel and uranium and everything else—was made of smaller particles, called atoms. But what atoms were made of no one knew. One common view was that they were something like tough and shiny ball bearings: mighty glowing entities that no one could see inside. It was only with the research of Ernest Rutherford, a great, booming bear of a man working at England's Manchester University, in the period around 1910, that anyone got a clearer view.

Rutherford was at Manchester, rather than at Oxford or Cambridge, not just because he was from rural New Zealand, and spoke with a common man's accent. If a research assistant was self-effacing enough, that could be overlooked. The problem rather was that when Rutherford had been a student at Cambridge he had refused to show proper deference to his superiors. He'd even suggested creating a joint-venture business to earn money from one of his inventions—and that was a mortal sin. Yet the reason he became the scientist who got the first clear glimpse of the inside of atoms was, to a large extent, because his heightened awareness of dis-

Ernest Rutherford

PHOTOGRAPH BY C. E. WYNN-WILLIAMS. AIP EMILIO SEGRÈ VISUAL ARCHIVES

crimination made him the kindest leader of men. The bluff booming exterior was just window dressing. He was good in nurturing skilled assistants, and his key experiment was monitored by a young man who would end up perfecting a most useful mobile radiation detection unit, of Rutherford's suggested design: the audibly clicking counter was to be Hans Geiger's claim to fame.

Their finding is so widely taught in schools today that it's hard to get back to the time when it was still surprising. What Rutherford realized was that these solid, impregnable atoms were almost entirely empty. Imagine that a meteor plummets into the Atlantic Ocean, but instead of staying down there, ultimately plonking against the seabed, we hear a great roaring, and see it come hurtling back out. Think how hard it would be to break through our preconceptions, and realize that the only way to explain it was that under the surface of the Atlantic there really wasn't smooth water all the way down. Rather, the analogy with what Rutherford had to deduce would be that the Atlantic's surface was just a thin liquid-rubbery film, and underneath it, where we had always thought there were deep waves and currents and tons of water, there was . . . Nothing.

It was all empty air, and a camera down there would show the arriving meteor, once it pierced the outer film, falling through empty space. Only at the very bottom, down on the sea floor, was there some powerful device, extremely compact, that could grab an incoming meteor, and send it hurtling up through the atmosphere, and back into outer space. The equivalent inside an atom is the atom's nucleus, lost far in the center. Only up near the outer surface of an atom are there the flurries of electrons that are involved in ordinary reactions, such as burning a piece of wood in a fire. But they're far from the central nucleus, which is shimmering deep below, within all the empty space.

If atoms were like little ball bearings, then Rutherford had found that these ball bearings were almost entirely hollow. There was just a tiny speck right at the center, called the nucleus. It was a disconcerting finding—the atoms we're composed of are mostly just empty space!—but by itself that still wouldn't have let anyone see how $E=mc^2$ could apply. The "solid" electrons up on the outer surfaces of the atom weren't going to pop out of their material existence and turn into exploding clouds of energy.

It was pretty clear that the nucleus was where scientists would have to turn next. There was a lot of electricity in the atom, and while half of it was spread diffusely, in the far-flung orbits of those electrons, the other half of it was crammed into the ultradense nucleus at the center. There was no known way to keep so much electricity concentrated in that small a volume. Yet something down there, in that nucleus, was able to squeeze down all that electricity, and hold it in a tight grasp, and keep it from squirmingly escaping. That must be where the storehouse—the hidden energy—that Einstein's equation hinted at could reside. There were positively charged particles—what we call the proton—in there, but no one could make out any greater detail.

An assistant of Rutherford's, James Chadwick, finally got an important better view, in 1932, when he detected yet another item locked inside the nucleus. This was the neutron, which got its name because although it roughly resembled the proton in size, it was electrically entirely neutral. It had taken Chadwick more than fifteen years to identify it. At one point students had put on a play about his quest for this particle that had so few properties it might as well, they teased, be called the "Fewtron." But if you've spent years putting up with Rutherford's booming impatience, you can handle students having their fun. Although Chadwick was a quiet man, he was pretty determined about what he would do.

Chadwick had originally been a slum boy from the Manchester streets, and his professional career had almost been destroyed just as it was about to begin. As a new postdoc under Rutherford, Chadwick had gone to Berlin, to study in the labs of the returned Hans Geiger. When World War I began, he meekly followed the advice of the local Thomas Cook's office that there was no reason to hasten to leave. As a result, he ended up spending four years as a POW, in the converted stables of a cold and windy Potsdam racecourse. He tried doing as much research as he could there, and even managed to get radioactive supplies. An enterprising firm, the Berlin Auer company, had extra thorium, and was marketing it to the German public in toothpaste as a way to make your teeth glow white. Chadwick simply ordered this miracle tooth whitener from the guards, then used it for his experiments. But he had such poor equipment that his tests never came to much. He was falling behind, and when he got back to England after the war ended in November 1918, barely managed to get back on track. Never again would he meekly follow anyone's advice.

In theory Chadwick's 1932 discovery of the neutron should have led immediately to further discoveries. A number of radioactive substances release neutrons, and those could be aimed like a submicroscopic machine gun at waiting atoms. Because neutrons were neutral, they wouldn't be bothered by the negatively charged electrons at the surface of the target atoms. When they reached the nucleus at the center, they shouldn't be bothered by any charges down there either. They'd be able to slip right in. Maybe you could use them as probes to see what was happening in there.

To Chadwick's disappointment, though, he could never get that to happen. The harder he blasted neutrons in at an atom, the less success he had in getting any of them to slip into the nucleus at the center. Only in 1934 did yet another researcher find a way around that problem, and manage to get neutrons to enter easily inside a target nucleus, to better see that nucleus's structure. And he wasn't working in an even more sophisticated research lab, but in the last place one might have expected.

The city of Rome, where Enrico Fermi lived, had memories of grandeur, but in the long decades leading up to the 1930s, it had steadily been left further and further behind the rest of Europe. The lab that the government gave Fermi, who was respected as one of Europe's leading physicists, was on an out-of-the-way street, in a quiet gardened park. There were tiled ceilings, and cool marble shelves, and a goldfish pond under the shady almond trees out back. For someone wanting to make a break from the mainstream European consensus, it was ideal.

What Fermi found in this gentle seclusion was that other research teams had been wrong to focus on blasting neutrons at higher and higher power to get them to enter the tiny nucleus inside an atom. Spraying fast neutrons directly at the great empty spaces inside a target atom meant that most of the neutrons simply raced right through. Only if the neutrons were *slowed*, so that they almost dawdled in their flight toward a distant nucleus, would they have a good chance of slipping inside. Slowed neutrons acted like sticky bullets. The reason they stuck so well to nuclei, one might visualize it, was that they became "spread out" in their relatively slow flight. Even if their main body missed the nucleus, the spread-out portions were still likely to connect.

On the afternoon when Fermi realized slowed neutrons could do this, his assistants lugged up buckets of water from the goldfish pond out back. They sprayed fast neutrons from their usual radioactive source into the water. The water molecules were of a size that made the incoming fast neutrons rebound back and forth till they slowed down. When the neutrons finally emerged, they were traveling slowly enough to slip regularly into any target nuclei ahead of them.

With Fermi's trick, scientists now had a probe that could get into the nucleus. But even that didn't make things entirely clear. For what was happening when the slowed neutrons entered? The full power that Einstein's equation spoke about still wasn't coming out. At most you got slightly changed forms of ordinary nuclei, which leaked out only a gentle sort of energy. It was useful for tracers that could be swallowed and then tracked to see what was going on inside the body. One of the first researchers to use similar tracers, George de Hevesy, employed it, his very first time, to prove that the "fresh" hash his Manchester boardinghouse landlady was serving was not quite as fresh as promised, but rather was coming back, slopped onto a fresh plate, steadily every day. But the slight energy leakages from elements that could be safely swallowed were not what the massive c^2 in the equation promised.

Somehow there had to be a further explanation; some further level of detail that physicists hadn't yet grasped. Atoms weren't solid massy spheres, but rather were almost entirely empty space—like an emptied ocean basin—with just the barest speck of a nucleus down at the center. That was what Rutherford had seen. The nucleus wasn't a simple solid speck either. It contained protons that crackled with positive electric charge, and pebblelike neutrons were packed in along with them. That was clear by 1932. The neutrons could go in and out of that nucleus pretty easily, once you did the unexpected twist of slowing them down when you sent them forward. That was what Fermi saw in 1934. But that's where matters stuck for several years.

$$E=mc^2$$

The solution to what was happening inside the nucleus—and so an unveiling of matter's deeper mechanisms, which would finally allow the energy promised by $E=mc^2$ to emerge—only came in 1938. It was provided by a solitary Austrian woman, sixty years old, stuck on the edge of Europe, in Stockholm; who didn't even speak Swedish.

"I have here . . ." she wrote, "no position that would entitle me to anything. Try to imagine what it would be like if . . . you had a room at an institute that wasn't your own, without any help, without any rights. . . ."

It was a dispiriting change, for just a few months earlier, Lise Meitner had been one of Germany's leading scientists—"our Madame Curie," as Einstein put it. She'd first arrived in Berlin in 1907, an impossibly shy student from Austria. But she'd tried to open up, and quickly became friends with one exceptionally good-looking young man at her university named Otto Hahn. He had an easygoing confidence, a self-teasing Frankfurt accent, and seemed to feel it a personal obligation to put this quiet newcomer at ease.

They were soon sharing a lab in the basement of the

Otto Hahn and Lise Meitner
CHURCHILL COLLEGE, CAMBRIDGE

chemistry department. They were almost exactly the same age, in their late twenties. He persuaded her to hum two-part harmony songs from Brahms with him, despite her off-key voice. When their shared work was going especially well, she wrote, "[Hahn] would whistle large sections of the Beethoven violin concerto, sometimes purposely changing the rhythm of the last movement just so he could laugh at my protests. . . ." The Physics Institute was nearby, and other young researchers there "often visited us and would occasionally climb in through the window of the carpentry shop, instead of taking the usual way." After working hours, Meitner remained solitary, living in a succession of single rooms, and sitting in the cheapest student seats at concerts she went to by herself. It was only at the lab that she found community.

She was a much better analyst and theoretician than Hahn, but he was bright enough—and sensible enough—to realize this would only be to his good; he had a history of finding excellent mentors. The first joint discoveries of Meitner and Hahn led to their getting a large lab in the new Kaiser Wilhelm Institutes, on what was then the western outskirts of Berlin. There were rural windmills still within sight; a forest a little farther to the west. They were becoming known as an important and trustworthy research team; they contributed to building up a core of indispensable knowledge about how atoms worked; their findings were soon as necessary to consider as those of Rutherford in England.

Through it all, she and Hahn kept their surface formality, carefully avoiding the informal "Du" form of address. In all her letters he was "Dear Herr Hahn." But there can be a special bond this way; a carefully unstated awareness that such dignified formality is blocking the pair off from any deeper links.

In 1912, after four years of working together, with Meitner now age thirty-four, Hahn married a younger art student. Meitner told everyone that it didn't matter. But although she'd never officially dated Hahn, she never dated anyone else in the years after that. There was another young colleague Meitner had been friendly with, James Franck, and she stayed in touch with him for over half a century, even when he got married, and then later when he was forced out of Germany to distant America. "I've fallen in love with you," Franck teased when they were both in their eighties. "*Spät!* (Late!)" Lise laughed.

In World War I, Meitner volunteered in hospitals, including some hellish ones near the eastern battlefields, while Hahn was on assignment with the army. The moral dilemmas of his work with poison gas seemed to bother neither of them. She sent letters regularly: lab gossip, and accounts of swimming trips with Hahn's wife, and occasionally the gentlest description of her hospital work. She also had a little time for research:

"Dear Herr Hahn! . . . Take a deep breath before you begin reading. . . . I wanted to finish some of the measurements so that I could . . . tell you a variety of delightful things."

Meitner had filled in one of the last gaps left in the periodic table listing all the elements. The work was her own, but she put both their names on it, and insisted to the *Physikalische Zeitschrift* editor that Hahn's name go first. During their wartime separations she tried not to push him for replies, but sometimes she slipped: "Dear Herr Hahn! . . . Be well, and write, at least about radioactivity. I remember a time very long ago when you would once in a while send a line even without radioactivity."

A little after the war they switched to different labs. By the mid-1920s Meitner headed the theoretical physics division within the Kaiser Wilhelm Institute for Chemistry. She was still shy on the outside, but had become confident in her intellectual work, regularly sitting in the front row with Einstein or the great Max Planck at the most respected theoretical seminars. Hahn was aware he couldn't follow such explorations, and cautiously stuck to more straightforward chemistry. But when Fermi's 1934 advances showed how the neutron might offer an ideal probing tool into the nucleus, Meitner shifted once again, to studies of the nucleus's properties. This meant she could hire Hahn, for chemists were always needed to study the new substances that were being formed.

In 1934 they started working together again, also taking on a recent doctoral student as their assistant, Fritz Strassmann. Hitler had come to power in 1933, but although Meitner was Jewish, and so immediately fired from the University of Berlin, she still was an Austrian citizen. The Kaiser Wilhelm Institutes had its own source of funding, and happily continued paying her as a full staff member.

But in 1938, Germany took over Austria, and Meitner became a German citizen by default. The institute might still be able to keep her on, but it would depend a lot on what her colleagues said. An organic chemist named Kurt Hess had long had a small office at the institute. He was a minor researcher, full of envy, and he was one of the first at the institute to become an active Nazi. "The Jewess endangers our institute," he began to whisper, to anyone who would listen. Meitner heard this from one of her ex-students, who had remained loyal. She talked it over with Hahn. Hahn went straight to Heinrich Hörlein, the treasurer of the organization that funded the Kaiser Wilhelm Institute for Chemistry.

And Hahn asked Hörlein to get rid of Meitner.

To say that people have been charming, as Hahn had been all his life, is simply to say that they've developed a reflex to do what will put the individuals around them at ease. It says nothing about their having a moral compass deeper than that. Hahn may have been slightly troubled by what he was doing to his old colleague: "Lise was very unhappy now that I had left her in the lurch." But most other German physicists did what the new government wanted them to, and many of Hahn's past students, pro-Nazi, were in positions of power as well. They—more than she—were the people he was increasingly working with now, the ones he needed to please.

He helped her a little bit with the details of leaving, but it's unclear how much Meitner understood in the shock. From her diary: "Hahn says I should not come to the Institute anymore. He has, in essence, thrown me out."

By the time she'd settled in Stockholm, in August 1938, Meitner didn't mention to anyone else what Hahn had done. Instead, almost by reflex, she just remained involved from a distance with the work she had been leading. With Strassmann and Hahn's help, she'd been guiding the streams of slowed neutrons into uranium, the heaviest of all naturally occurring elements. Since neutrons slipped into and then stuck within the nuclei they hit, everyone expected that the result would be some new substance, even heavier in weight than the uranium they started with. But try as she and the researchers in Berlin might, they couldn't clearly identify whatever new substances they were creating.

Hahn, as ever, seemed the slowest to grasp what was happening. Meitner met him, in neutral Copenhagen in November, and after he admitted he didn't have a clue, she sent him back with clear instructions for more experiments. He just had to use the top-quality neutron sources and counters and amplifiers she'd assembled, and which were still in place in their lab, right where she'd left them. The mail was so quick between Stockholm and Berlin that she could even talk him through the steps. "Meitner's opinion and judgment carried so much weight with us in Berlin," Strassmann recounted later, "that we immediately undertook the necessary . . . experiments." However much she was wounded, at least she could continue with the work that had been her focus for years.

Meitner suggested they keep an eye out for variants of radium that might be produced in the long bombardment process that had started with uranium. (Radium is a metal with a nucleus almost as massive as that of uranium. Both are so overstuffed with neutrons that they regularly end up spraying out radiation.) At this stage it was just a hunch, based on similarities between the two metals, and the fact that they were so often found together in mines.

But it meant that the broader effects of $E=mc^2$ were, finally, about to appear.

> Monday evening in the lab
>
> Dear Lise!
>
> . . . There is something about the "radium isotopes" that is so remarkable that for now we are telling only you. . . . Perhaps you can suggest some fantastic explanation. . . . If there is anything you could propose that you could publish, then it would still in a way be work by the three of us!
>
> Otto Hahn

They had been using ordinary barium as something of an adhesive in the lab, to gather the fragments of neutron-loaded radium. Once the barium had done that job, it was collected with acids and then rinsed away. The problem now, though, was that Hahn could not get it to separate. Some of the barium that was left always seemed to have tiny bits of something radioactive stuck to it.

He and Strassmann were at a loss. "Meitner was the intellectual leader of our team," Strassmann explained. But now she wasn't here. Hahn wrote her again, two days later: "You see, you will do a good deed if you can find a way out of this." They could do no more. The strange result—why couldn't they get the radiation away from the simple barium?—would be up to her to try to work out.

It was nearly Christmas by this time, and a couple who knew that Meitner was alone in Stockholm invited her to stay at a hotel in their vacation village of Kungälv, on the west coast of Sweden. A nephew of hers whom she'd always liked, Robert Frisch, was in Copenhagen, and on Meitner's suggestion, the couple invited him too.

Meitner had first really come to know her nephew when he'd been an eager science student in Berlin a decade before. They'd often played piano duets together, even though she had trouble keeping up. (Though they'd have fun, translating *Allegro ma non tanto* as "Fast, but not auntie.")

Now Robert was a grown man, and a promising physicist, working at Niels Bohr's institute in Denmark. The first night, arriving late, he was in no condition for discussing science. The next morning, when he came down to the ground-floor restaurant at their hotel, he found his aunt puzzling over Hahn's letter. The barium they had added was showing such persistent radioactivity—so much spraying out of energy streams—that she and the researchers in Berlin couldn't help but wonder why. Had it somehow been created that way during the Berlin experiments?

Frisch suggested that it was just a mistake in Hahn's experiment, but his aunt waved that aside. Hahn was no genius, but he was a good chemist. Other labs made mistakes. Not hers. Frisch didn't take much convincing. He knew she was right.

They stayed at the breakfast table while Frisch ate, talking it over. The experiment that Meitner had suggested to the Berlin crew could be explained if the uranium atom had somehow split apart. A barium nucleus is about half the size of a uranium nucleus. What if the barium they were detecting was simply one of the big halves that had resulted? But by everything nuclear physics had been showing—all the work from Rutherford on up—that should be impossible. There are over

200 particles inside a uranium nucleus, all those protons and neutrons. They were stuck together with what is known as the strong nuclear force, an exceptionally powerful nuclear glue. How could a single incoming neutron break through every one of those bonds, to tear off a huge chunk? You don't throw a simple pebble at a large boulder and expect the boulder to break in half.

They finished breakfast, then went for a walk in the snow. Their hotel wasn't far from a forest. Frisch put on skis, and offered to help his aunt with a pair for herself, but she declined ("Lise Meitner made good her claim," Frisch wrote, "that she could walk just as fast without").

No one had ever chipped off more than a fragment from a nucleus. They were confused. Even if an incoming neutron had hit some sort of weak point, how could dozens of protons be pulled off in one impact? The nucleus wasn't built like a rocky cliff that could break in half. It was supposed to remain intact for billions of years.

Where could the energy to suddenly tear it apart come from?

Meitner had first met Einstein at a conference in Salzburg, in 1909. They were almost exactly the same age, and Einstein was already famous. At the conference he recapped his 1905 findings. To find that energy could appear out of disappearing mass was "so overwhelmingly new and surprising," Meitner recounted decades later, "that to this day I remember the lecture very well."

Now, in the snow with her nephew, she stopped by a tree trunk, and they settled down to work it out. The most recent model of the nucleus was due to Niels Bohr, the kindly soft-spoken Dane who was her nephew's employer. Instead of looking at the nucleus as a rigid metal, some stiffened collection of ball bearings welded tight, Bohr viewed it more as a liquid drop.

A water drop is always on the verge of bursting apart, due to the weight inside it. That near-bursting weight is like the crackling electric charges between the protons in a nucleus. All the protons push against each other. (That's what two positive charges will always do.) But a water drop stays together, most of the time, because it also has a lot of rubbery surface tension on the top. That is like the glue-taut strong force that clusters the protons together, despite all the electricity trying to push them apart.

In a small nucleus, such as that of carbon or lead, the gluing strong force is so great that it doesn't matter that there's a lot of electrical power hidden away inside, trying to push the protons apart. It won't win. But in a big nucleus, a really huge one such as that of uranium, could the extra neutrons tip the balance?

Meitner and her nephew weren't physicists for nothing. They had paper with them, and pencils, and in the cold of the Swedish forest, this Christmas Eve, they took them out and began calculating. What if it turned out that the uranium nucleus was so big, and so crammed with extra neutrons spacing apart the protons in there, that even before you started artificially pushing extra neutrons in, it was already in a pretty precarious state? That would be as if the uranium nucleus were a water droplet that already was stretched apart as far as it could go before bursting. Into that overstuffed nucleus, one more plump neutron was then inserted.

Meitner started to draw the wobbles. She drew as well as she played the piano. Frisch took a pencil from her, politely, and did the sketches. The single extra neutron that came in made the nucleus begin to stretch in the middle. It was like taking a water balloon, and squeezing it in the middle. The two ends bulge out. If you're lucky, the rubber of the balloon will hold, and the water won't burst out. But keep on with it. Squeeze in some more, and when the balloon spreads sideways, let go until it rebounds back toward the center and then squeeze in the opposite way. Keep on repeating. Eventually the balloon will burst. Get your timing right, and you don't even have to squeeze very hard. Each time the water balloon is rebounding back, you just let it reach its maximum rebound, and then—as with pumping on a swing—you give it a further squeeze to speed it on its way into yet another rubber-stretching contortion.

In the uranium nucleus, that's what the incoming neutrons had been doing. The reason Hahn had so much trouble classifying what he saw was that he'd been convinced adding neutrons would only make a substance heavier. But, in fact, he'd cracked the uranium in half.

It was a crucial insight, if true, but they'd have to check it. To start with, they knew that the electricity of the protons within the nucleus could now be available to make the bits fly apart. In the units by which physicists keep count, that's about 200 MeV—200 million electron volts. Frisch and Meitner worked that out mostly in their heads. But would Einstein's 1905 equation prove that there really was that amount of energy available inside to send the nucleus roaring apart? Frisch takes up the story:

Fortunately [my aunt] remembered how to compute the masses of nuclei . . . and in that way she worked out that the two nuclei formed by the division of a uranium nucleus would be lighter than the original uranium nucleus, by about one-fifth the mass of a proton. Now whenever mass disappears, energy is created, according to Einstein's formula $E=mc^2$. . . .

But how much energy would that be? One-fifth of a proton is a preposterously tiny speck of matter. The dot over a letter *i* has many more protons than there are stars in our galaxy. Yet the "disappearance" of that fifth of a proton—this subvisible speck—has to be enough to generate 200 MeV of energy. In Berkeley, California, a building-sized magnet was being planned that might, when charged with more electricity than the whole city of Berkeley ordinarily used, power up a particle to 100 MeV of energy. And now this speck was supposed to produce even more.

It would seem impossible—except for the immense size of c^2. The world of mass, and the world of energy, are linked by that frantically widening bridge. From our perspective, the fragment of a proton slips across the roadway of that "=" sign: transforming; growing.

Growing.

They had crossed a river on their walk out from Kungälv, but it was frozen. The village was too far away to hear any market noises. Meitner did the calculation. Frisch remembered later: "One-fifth of a proton mass was just equivalent to 200 MeV. So here was the source for that energy; it all fitted!"

The atom was open. Everyone had been wrong before. The way in wasn't by blasting harder and harder fragments at it. One woman, and her nephew, quiet in the midday snow, had now seen that. You didn't even have to supply the power for a uranium atom to explode. Just get enough extra neutrons in there to start it off. Then it would start jiggling, more and more wildly, until the strong forces that held it together gave way, and the electricity inside made the fragments fly apart. This explosion powered itself.

Meitner and her nephew considered their science politically neutral, so they prepared their discovery for publication. It had to be named, and Frisch was re-minded of how bacteria divide. Back in Copenhagen he asked a visiting American biologist at Bohr's institute for the right word in English. The label *fission*, accord-ingly, to describe how atomic nuclei divide, was intro-duced in his ensuing paper. Hahn had already published the Berlin findings, with minimal credit to Meitner, and soon began a nearly quarter-century-long campaign to pretend that all the credit was really his own.

The thirty-year quest was over. In the decades since Ein-stein's equation first appeared in 1905, physicists had shown how the atom could be opened, and the com-pressed and frozen energy that $E=mc^2$ spoke about let out. They'd found the nucleus, and a particle called the neutron that could get in and out of the nucleus pretty easily (especially if one used the trick of sending it in slowly), and they'd found that when extra neutrons were pushed into overstuffed atoms such as uranium, the whole nucleus wobbled, and trembled, and then ex-ploded.

What Meitner had realized was that this could occur because of the way the powerful electricity within the nucleus was held in by the springs or glue of the strong nuclear force. When an extra neutron made the nucleus start wobbling, those springs gave way, and the inner parts flew apart with wild energy. If you checked all the weights before and after, you'd find that the bits flying apart "weighed" less than when they were still held to-gether in the nucleus. That "disappeared" mass was what powered their high-velocity escape. For it hadn't truly disappeared. The deep insights behind the equa-tion had guaranteed that it would simply become ap-parent in the form of energy, getting the powerful c^2

boost to magnify it (in units of mph^2) by nearly 450,000,000,000,000,000 times.

It was an ominous finding, for in theory anyone could use it to start cracking apart nuclei, those central cores of atoms, and letting out these great blasts of energy. In any other era, the next steps might have taken place slowly, with the first atomic bomb only appearing some time in the 1960s or 1970s. But in 1939, the world had just begun its largest war ever.

The race was on to see in which country the equation's power would emerge first.

PART 4

Adulthood

10

$$E=mc^2$$

By 1939, Einstein was far from being the unknown young man whose father had to beg a Leipzig professor to give him a job. His work in relativity had made him the most famous scientist in the world. He had been Berlin's leading professor, and when anti-Jewish mobs and politicians had made it impossible to stay there, he went to America, in 1933, taking up a position at the new Institute for Advanced Study in Princeton, New Jersey.

When Einstein heard of Meitner's results and how other research teams were beginning to extend it, colleagues were able to get one of the president's own confidants to carry a personal letter direct to the White House.

> F. D. Roosevelt,
> President of the United States,
> White House
> Washington, D.C.
>
> Sir:
>
> Some recent work . . . which has been communicated to me in manuscript, leads me to expect that the element uranium may be turned into a new and

important source of energy in the immediate future. Certain aspects of the situation which has arisen seem to call for watchfulness and, if necessary, quick action on the part of the Administration. . . .

This new phenomenon would . . . lead to the construction of bombs, and it is conceivable—though much less certain—that extremely powerful bombs of a new type may thus be constructed. A single bomb of this type, carried by boat and exploded in a port, might well destroy the whole port together with some of the surrounding territory. . . .

Yours very truly,
Albert Einstein

Unfortunately, it met with this reply:

THE WHITE HOUSE
WASHINGTON

October 19, 1939

My dear Professor,

I want to thank you for your recent letter and the most interesting and important enclosure.

I found this data of such import that I have convened a board. . . . Please accept my sincere thanks.

Very sincerely yours,
Franklin Roosevelt

Even someone who'd only been in America for a few years, as Einstein had, would understand that "most interesting" was a brush-off. Presidents are constantly sent

Albert Einstein
IMAGE SELECT/ART RESOURCE, NEW YORK

impractical ideas. There's an obligation to be polite when the sender is famous, but FDR and his colleagues did not believe a bomb could possibly destroy a whole port.

The letter was shuttled away from FDR's desk, and ended up in the hands of Lyman J. Briggs, the easygoing, pipe-smoking director of the federal government's Bureau of Standards. He would be responsible for all U.S. atomic bomb development.

In the long history of governments assigning the wrong man to a job—and there have been some choice ones—this is one of the choicest. Briggs had entered government service during the administration of Grover Cleveland, in 1897, before the Spanish-American war. He was a man of the past, comfortable with that time when everything had seemed easier, and America had been safe. He wanted to keep it that way.

In April 1940, Meitner's nephew, Robert Frisch, then in England, was beginning to convince the British authorities that a practical bomb could be built. A top-secret memo carrying the news was later rushed to

Washington. By then there had been massive battles throughout Europe; panzer armies had overrun ever more countries. But you couldn't trick Lyman J. Briggs. That darn-foolish Brit report could be a danger if it ever got out. He locked it in his safe.

Germany's bureaucrats, even scientifically untrained ones, took the opposite view of history. What good was the recent past? It had only led to the sellout at the end of World War I, the corruption of the Weimar Republic, inflation, and then unemployment. The beckoning future would be better. That's why there was such belief in new roads, new cars, new machines—and new conquests. The latest laboratory speculations also promised something new and powerful. Joseph Goebbels later noted in his diary: "I received a report about the latest developments in German science. Research in the realm of atomic destruction has now proceeded to a point where . . . tremendous destruction, it is claimed, can be wrought with a minimum of effort. . . . It is essential that we be ahead of everybody. . . ."

And they had just the man for the job.

In the summer of 1937, early in the month of July, Werner Heisenberg was on top of the world. He was the world's greatest living physicist after Einstein, famous for his work in quantum mechanics and the Uncertainty Principle. He had just been married, and now was returning after an extended honeymoon to the old family apartment in Hamburg, where his mother still lived, and the old five-foot-long electrically operated battleship he'd made as a teenager was still on display. He had a pleasant phone call to make, for he'd also been appointed to a senior position at the same university department where he'd earned his own Ph.D., as the wonder of the German academic establishment almost fifteen years earlier. He dialed the university's rector from his mother's phone.

Heisenberg had a way of standing with shoulders squared straight, in a state of alert excitement, whenever he was pleased. The call went through, but the rector told him there was a serious problem. An elderly physicist, Johannes Stark, had convinced the weekly magazine of the SS to run an anonymous article saying that Heisenberg wasn't sufficiently patriotic, that he had worked with the Jews, didn't have the proper pro-German spirit, etc.

This was the sort of public attack that often preceded a late-night arrest and then deportation to a concentration camp. Heisenberg was scared, but also furious. They were picking on the wrong man! It's true he'd worked with Jewish physicists, but Bohr and Einstein and the great physicist Wolfgang Pauli and so many others were Jewish or partly Jewish that he'd had no choice. Despite that he'd always stood up for his country in public discussions, defending Hitler's actions; he'd always faithfully rejected job offers from top foreign universities.

Heisenberg tried enlisting closest friends to help, but that had no effect. Soon he was brought for questioning to the basement of SS headquarters at Prinz-Albert-Strasse in Berlin, where the walls were uncovered cement, and the mocking sign "Breathe deeply and calmly" was up. (He wasn't beaten, and one of the interrogators had taken a Ph.D. at Leipzig for which Heisenberg was an examiner, but his wife later said he had nightmares about it for years.) Only when there were no signs of the SS attack letting up did he enlist one more ally, the woman who was closest to him of all: his mother.

The Heisenbergs were from the educated middle class, and so were the Himmlers, and Heisenberg's mother had known Heinrich Himmler's mother from the time they were young. In August, Mrs. Heisenberg went to see Mrs. Himmler, in her small but very clean apartment, where fresh flowers were always placed in front of the crucifix, and she passed along a letter from her son.

At first Mrs. Himmler didn't want to bother *her* son by delivering it, but as Heisenberg later recalled, his mother played the trump card: " 'Oh, you know, Mrs. Himmler, we mothers know nothing about politics—neither your son's nor mine. But we know that we have to care for our boys. That is why I have come to you.' And she understood that."

It worked.

[From the office of the director of the SS]

Very Esteemed Herr Professor Heisenberg!

Only today can I answer your letter of July 21, 1937, in which you direct yourself to me because of the article of Professor Stark. . . .

Because you were recommended by my family I have had your case investigated with special care and precision.

I am glad that I can now inform you that I do not approve of the attack . . . and that I have taken measures against any further attack against you.

I hope I shall see you in Berlin in the fall, in November or December, so that we may talk things over thoroughly man to man.

With friendly greetings,

Heil Hitler!

Yours,

H. Himmler

P.S. I consider it, however, best if in the future you make a distinction for your audience between the

results of scientific research and the personal and
political attitude of the scientists involved.

The P.S., it was understood, meant that Heisenberg
could use Einstein's results on relativity and E=mc², but
he had to show he disavowed Einstein himself, and gave
no support to the sort of liberal or internationalist
views—such as supporting the League of Nations or
speaking out against racism—that Einstein and other
Jewish physicists had been known for taking.

These terms weren't hard for Heisenberg to accept.
In his teens he'd been active in the wandering clubs,
where young Germans hiked for days or weeks at a time
through wilderness areas, getting in contact with the
true roots of their nation. Often, around the campfire,
they'd talk of mystical heroes, and the way their country
might regenerate in a hoped-for "Third" Reich, to be
led—in the common phrasing of the time—by a sufficiently far-seeing leader, or *Fuehrer*. Many youngsters
grew out of this, but well into his twenties Heisenberg
remained drawn to the movement, despite mocking remarks from his more grown-up or liberal colleagues.
During advanced university studies he'd regularly leave
a seminar and meet up with a group of teenage boys, to
lead them for a long walk and if possible a full night out
in the woods, before racing back, by train, just in time
for his next 9:00 A.M. lecture.

When the German army's Weapons Bureau began
work, in September, 1939—a month after Einstein's letter to FDR—Heisenberg was one of the first to volunteer
to do anything that was needed. The Reich was already
at war: its artillery and ground troops and air force and
panzers successful in Poland. But there might well be
greater enemies still to come. Heisenberg had always
been an energetic worker, and now he surpassed him-

self. In December he delivered the first part of a comprehensive paper on how to construct a workable atomic bomb. In February, 1940, he had the full report done, and when a Berlin unit was set up to start building a reactor, in parallel with the work at his home university of Leipzig, he took command of both, regularly commuting between the two cities. It would have been tiring for most men, but Heisenberg was at the peak of his powers. He was still only in his thirties, a regular mountain-climber, a rugged horse rider, and active with two months' training each year in a mountain infantry unit, facing the Austrian border.

The first tests were in Berlin, in an ordinary plank-walled building built on the wooded grounds not far from Meitner's old institute, under the cherry trees that bloomed so well in the warm, clear summer of 1940. To keep out the curious, the building was named "The Virus House." Heisenberg's first step was to load in enough uranium—far more than the fractions of an ounce Meitner had arranged for Hahn to try in 1938. Only a very few atoms had exploded apart then. The sample had been so thin that most of the neutrons that the atoms released simply sped off into the lab's empty air.

Heisenberg ordered dozens of pounds of uranium. It wasn't hard to get, because the Reich's army had taken over Czechoslovakia a full year before invading Poland. Europe's largest uranium mines, which Marie Curie had once used, were at Joachimsthal there. The uranium was delivered. With the prestige of Heisenberg's name, the Weapons Bureau could ensure there were priority train shipments.

Simply stacking together a large amount of uranium wouldn't get a reaction going. For a nucleus, as we saw, is a vanishingly tiny speck, deep in the empty spaces in-

Werner Heisenberg (seated third from right). He's wearing an armband in mourning for his father.

AIP EMILIO SEGRÈ VISUAL ARCHIVES, PEIERLS COLLECTION

side an atom. Most of the neutrons released by the first explosions would speed right past the nucleus, like an alien space probe hurtling through our solar system.

The twist Fermi had found—that there was a great power in using slowed neutrons—could help solve this problem, and get a reaction to begin. Fast neutrons can be thought of as sleek, "streamlined" as they travel. But slow neutrons, as we've seen, can be thought of as more "wobbly"; more spread out. Even if their main body only comes close to one of the waiting nuclei, then part of the neutron—the "spread-out bit"—is likely to connect. What would have been a near miss of a nucleus if the neutron were coming in fast, becomes far more likely to be a "catch." When such a slow neutron gets caught or pulled in, circumstances are prepared for $E=mc^2$ suddenly to operate: for the nucleus to wobble, then explode, letting out its great blasts of energy, and also, in the process, letting out a gush of extra neutrons, so that yet more atoms are hit, to wobble and explode apart in turn.

Heisenberg searched for the right substance to pro-

125

duce that useful slowing of the neutrons. Anything that was dense with hydrogen would work to some extent, because the bumpings of neutrons against hydrogen tends to slow the neutrons down. That was why Fermi, back in 1934, had managed to get some effect even with ordinary water (H_2O) from the goldfish pond on the grounds of his institute. But when the first German teams tried spreading ordinary water around a uranium sample, although there were a few crackling reactions at the center of the sample, where the first uranium atoms broke apart, the neutrons that gushed out were still moving too fast to get the reaction to spread.

Heisenberg needed a better moderator. He knew that around the time when Fermi had been working, a U.S. chemist, Harold C. Urey, had discovered that the water in all the world's oceans and lakes is not composed merely of H_2O. Mixed in with it is a variant molecule, slightly heavier. Instead of having ordinary hydrogen at the core, it has deuterium, which is very similar to hydrogen but weighs about twice as much. Aside from that it is the same as ordinary water—it's just as free-flowing, and transparent; it's part of our rain and ice and seas; we drink it all the time. But there's only one molecule of it for every 10,000 molecules of ordinary water, which is why "heavy water" had been overlooked before. (A large swimming pool has only about a drinking glass's worth spread within it.) But heavy water is superb at slowing down high-velocity neutrons. They smack into the heavier deuterium and start ricocheting, slowing with each bounce, and emerge a fraction of a second later, after several dozen ricochets, floating much slower than when they went in.

German labs had accumulated only a few gallons of heavy water at first. This wasn't enough to be shared between Leipzig and Berlin. Heisenberg's sentiments were more with Leipzig, so it was in the basement of the physics institute there that the most important tests were prepared. As 1940 went on, the precious heavy water was poured in with the several pounds of stacked uranium. The mix of heavy water and uranium was packed into a tough spherical containment vessel, then dangled from a hoist while measuring devices were set up around it. Professors weren't supposed to get involved with the minutiae of experiments, but Heisenberg prided himself on his practical skills as much as on his theoretical gifts. He had built some of those measuring devices himself, with Robert Döpel, the main experimenter in charge.

When the uranium and heavy water were in place, and the measuring devices set up closely around the container, the experiment was ready to begin. Gunpowder needs a match to start. Dynamite needs a blasting cap. For an atomic reaction, even one that's in uranium of too low a quality to lead to a full explosion, there has to be an initial source of emerging neutrons. Döpel had left a hole in the bottom of the containment vessel. The neutron source was a small amount of radioactive substance, similar to the one Chadwick had used. It was brought over, contained in a single long probe, and now, for the first time—in an experiment that began in February 1941—all the working parts that could come together in a bomb were in place.

When Döpel and Heisenberg gave the instructions, the probe would go in, and the first speeding neutrons would be let loose inside the uranium. A few of the uranium atoms would burst, flying apart much faster than anyone would have suspected before Meitner explained how $E=mc^2$ was operating. Even more neutrons would spray out in the fast debris. They'd pass without much effect through the first layers of uranium atoms they hit, but when they reached the heavy water they'd be caught bouncing back and forth till they slowed down. When they reached the next layer of uranium metal they'd be wobbling so slowly, and so dispersedly, that they would be much more likely now to connect with the uranium nuclei, and especially the most fragile ones, and overload them, making them wobble, and tremble, and then explode apart in turn.

Each one would be another crackling occurrence of $E=mc^2$, in a sequence that Geiger counters would show building up faster and faster. In the first few millionths of a second—by Heisenberg's calculations—there would be perhaps 2,000 explosions. In the next millionths of a second there would be 4,000. Then 8,000, then 16,000, and so on. Doubling on that time scale rushes upward very quickly. If everything worked, there would be trillions of these minute explosions still in only a fraction of a second, and then hundreds of trillions, and the cascading effect would keep on going. It would be a "rip" in the ordinary fabric of matter, as all the energy that had been squeezed inside these atoms for billions of years now came out: there in the basement of the Leipzig institute; in this university run by officers appointed from Reich headquarters; with students proudly wearing the swastika in the classrooms above. To explode apart a billion atoms, you wouldn't need to set up an enormous laboratory, with a billion initiating neutron machines. Once a very few atoms started, the debris fragments they sent out—loaded with neutrons—would quickly start up the rest. This first uranium wasn't purified enough to produce a runaway reaction, but it would be a start.

The professors gave the instruction. Döpel's assistant, Wilhelm Paschen, inserted the probe. It was early 1941. The initiating neutrons were inside the uranium! Everyone stared at the dials to record the result.

And nothing happened.

There wasn't enough uranium to get a reaction going. Heisenberg wasn't fazed, and simply ordered more, from the enterprising Berlin Auer company, which over the years had moved on from toothpaste, and was now a wholesale supplier of a whole range of uranium products. Its raw supplies were no problem, as Einstein had also warned in his letter to FDR. ("Germany has actually stopped the sale of uranium from the Czechoslovakia mines which she has taken over," Einstein had written, ". . . while the most important source of uranium is the Belgian Congo.") The Union Minière in occupied Belgium had thousands of pounds stockpiled from those Congo mines. When the Joachimsthal stockpiles ran low, the Germans turned there next.

Machining uranium into a usable form was hard, since it demands a lot of labor, plus the fine uranium dust that's produced is dangerous for the workers. But Heisenberg had a procurement organization to draw on that wasn't hindered by outmoded notions of human rights. Germany had many concentration camps, full of people who were soon to be killed anyway. Why shouldn't important projects get some advantage from them? As the war went on, Berlin Auer executives calmly bought female "slaves" from the Sachsenhausen concentration camp. They could be worked to prepare the uranium oxide the German project needed. Back in April 1940, Heisenberg had expressed his impatience at how long the first Auer shipments were taking. The administrative head of the army project assured him that Berlin Auer would have its work force operating at maximum speed. Supplies had started coming in that summer, and now in 1941, even more was quickly shipped.

In autumn 1941 there were more promising tests,

and finally, in spring 1942, the breakthrough happened. The containment vessel was pouring out neutrons: 13 percent more than the inserted source had pumped inside it to start with. The trapped energy that Einstein had first spoken of nearly 40 years before was now being released. It was as if a narrow funnel was stretching up from deep underground, and a thundering wind—the released energy—was blowing fast along it. Himmler's faith was justified. Heisenberg—triumphant—had managed to get the power of Einstein's equation to come roaring up, and appear in Nazi Germany.

Einstein was getting hints of Heisenberg's success, for the director of the Kaiser Wilhelm Institute for Physics had been Dutch, and when he too was expelled, ending up in America, he shared with his new colleagues what he'd heard of the work at the Virus House and Leipzig.

Einstein wrote another letter to FDR: "I have now learned that research in Germany is carried out in great secrecy and that it has been extended to another of the Kaiser Wilhelm Institutes, the Institute of Physics." But this time it seems that there wasn't even the courtesy of a reply. A white-haired foreigner is one thing, especially if he has a grand reputation in science. But tensions were rising as war got closer, and the FBI now saw reasons to discount anything he said. For Einstein was a socialist, and a Zionist—and he had even spoken out against excess profits for arms manufacturers. The FBI reported back to army intelligence that:

> In view of his radical background, this office would not recommend the employment of Dr. Einstein, on matters of a secret nature, without a very careful investigation, as it seems unlikely that a man of his

background could, in such a short time, become a loyal American citizen.

When the United States finally did get a serious atomic project started, it was helped through some skillful manipulations by impatient visitors from Britain. Mark Oliphant was another one of Rutherford's bright young men, and in the summer of 1941 he led a two-front assault. First he arrived in Washington, dangling the gift of the cavity magnetron—a key device for shrinking room-sized radar sets to a volume that could be crammed into an airplane, and also for greatly increasing accuracy. (This was when Oliphant discovered that Lyman J. Briggs, leader of the West's atomic research project, had locked the top-secret British results inside his safe.) Next, Oliphant traveled to Berkeley, where the physicist Ernest Lawrence worked.

Lawrence was not especially bright as physicists go, but he loved machines, great big powerful machines, and his very simplicity—his directness of focus—allowed him to get them built. For example, Samuel Allison (working at the University of Chicago then) remembers that Briggs had "a tiny cube of uranium which he liked to keep on his desk and show to insiders. . . . Briggs used to say, 'I want a whole pound of this,' . . . Lawrence would have said he wanted forty tons and got it."

By the autumn of 1941 Briggs was out, and a group of more effective leaders including Lawrence was in, and by December—when Pearl Harbor brought the United States into the war—the project really took off. It came to be called the Manhattan Project, as part of the cover story that it was simply part of the Manhattan Engineering District.

The refugees Briggs had scorned were indispensable. Eugene Wigner, for example, was a remarkably quiet, unassuming young Hungarian, who came from an equally quiet and unassuming family. When World War I had broken out, Eugene's father had stayed away from political discussions, pointing out, quite sensibly, that he was pretty sure the emperor was not going to be swayed by the views of the Wigner family. But this caution meant that when Eugene, a superb student, was facing university choices, the father had him take a practical engineering degree, as the odds on a career in theoretical physics succeeding were very slight.

Wigner did succeed at physics, and after he was forced out of Europe in the 1930s, he ended up centrally involved in the American duplicate of Heisenberg's calculations, detailing how a reaction could begin. But his engineering training meant that he handled the subsequent steps far better than Heisenberg. What shape, for example, should the uranium be that would go inside a reactor? The most efficient possible design would be a sphere. That way the maximum number of neutrons would be deep in the center. Next best—if a sphere was too hard to cut accurately—would be an oval shape. After that comes a cylinder, then a cube, and last, worst of all, would be to try building it with the uranium stretched out in flat sheets.

For his Leipzig device, Heisenberg had chosen the flat sheets. The reason was simply that flat surfaces almost always have the easiest properties to compute, if you're advancing by pure theory. But engineers with enough practical experience are never restricted to pure theory. There are many informal tricks of the trade for how ovals and other shapes work. Wigner knew them, as did many other similarly cautious refugees, who'd also been advised by their families to take engineering degrees. Heisenberg did not. That was of central importance. Professors in general tend to be hierarchical, and pre-World War II German professors were at the peak of such confidence. As the war went on, a number of junior researchers in Germany found that Heisenberg had been mistaken in one engineering assumption or another. But Heisenberg almost always refused to listen; would angrily try to keep them from even daring to mention it.

Even so, nobody could be confident the United States was going to win the race to make the bomb. America was just coming out of the Great Depression; much of its industrial base was still rusted and abandoned. When Heisenberg began his research for army ordnance, the Wehrmacht was the world's most powerful fighting force. It had entire army groups supplied with equipment that surpassed that of any other nation. The United States had an army that, if you included a lot of generation-old World War I artifacts, could just about supply two divisions, thus placing it below the tenth rank in the world, at about the level of Belgium.

Germany also had the world's best engineers, and a strong university system—despite having expelled so many Jews—and above all, they had that head start: two precious years when Heisenberg and his colleagues had been working full out, while Briggs had mostly been musing at his desk. These were the quirks of fate that would influence who ended up using the equation first. $E=mc^2$ was far from the pure reaches of Einstein's inked symbols now. The Allied effort would have to go faster.

The German effort would have to be sabotaged.

E=mc²

British intelligence had been monitoring the German program from the beginning, and identified its one weak spot. It was not the uranium—there was too much in Belgium to try to destroy, even if anyone could get at it. Nor was it Heisenberg himself: no assassination squad could reach him in Berlin or Leipzig, and even his family's summer resort in the Bavarian Alps was too far away, and probably too well guarded as well.

The most vulnerable target, rather, was the heavy water. A reactor couldn't fully ignite without slowing down the neutrons from the first atom explosions, so they could find the other nucleus specks, and jostle them to start exploding their hidden energy in turn. Heisenberg had decided on heavy water for that, but it takes a very large factory—using a great amount of energy—to separate that from ordinary water.

Some cautious members of Heisenberg's staff had proposed that Germany construct a factory of its own, safely on German soil, to produce the heavy water. But Heisenberg, backed by army officials, knew there already was a perfectly sound heavy water factory in operation, using the abundant, power-generating waterfalls of

Norway. It's true that until recently Norway had been an independent country, but now wasn't it merely a conquered province?

It was a fateful decision, but generations of German nationalists had felt their country was suffocated, entrapped. Heisenberg backed the decision to rely on the Norwegian factory, for he backed the idea of the new Reich's right to dominate all of Europe. Through the war he excitedly visited one subject nation after another, striding through the offices of his onetime colleagues, local collaborators nearby; in the Netherlands explaining to the aghast Hendrik Casimir that although he knew about the concentration camps, "democracy can't develop sufficient energy," and he "wanted Germany to rule."

The Norwegian factory was located up a mountainous ravine, at Vemork, 90 miles by winding road from Oslo. Before the war it had produced only 3 gallons—24 pounds—of heavy water a month, for laboratory research. Engineers from the great I. G. Farben industrial combine in Germany had asked for more, and offered to pay above market rates, but the Norwegian managers had refused, unwilling to help Nazis. A few months later the Farben engineers asked again; this time—for the Wehrmacht had destroyed the Norwegian army—they were backed by troops with machine guns. The Vemork staff had no choice but to agree. Production had been accelerated to an annual rate of 3,000 pounds by mid-1941. Now, in mid-1942, it was up to 10,000 pounds per year, steadily shipped to Leipzig, Berlin, and the other centers for atomic research.

There were only a few hundred troops guarding the factory, for the site seemed impregnable. The Norwegian resistance was clearly too small and untrained to be feared for an assault on such a huge factory. The com-

The Norsk Hydro plant at Vemork, Norway
NORSK HYDRO

plex was surrounded by barbed wire and arc lights, with only a single suspension bridge giving access. It was located in a setting so deeply cut in the mountains that for over five months of the year shadows from the surrounding peaks kept direct sunlight entirely away, and workers had to be taken up by cable car, to a higher plateau, to get a daily dose of light.

This was the target the British government chose to attack. If Vemork had been on the coast, then members of the Royal Marines could have tried to go in, but since it was 100 miles inland, a team from the First Airborne Division was chosen. These troops were good. Many were working-class London boys, fists trained from surviving the Depression, and now in their twenties, undergoing more serious training: weapons, radios, explosives. They weren't told where they were going, of course—that would only come on the day of the mission. Until then they believed they were being prepared for a paratrooper competition with the Yanks. That their fate was being directed in an effort to control what Einstein's equation and Rutherford's investigations were leading to—of that they had no idea at all.

Two glider teams took off after dark from northern Scotland, towed by the new high-speed Halifax bombers. There were about thirty troops in all. (Today we think of a typical glider as a single-man device, but then, before helicopters were widespread, they were often much bigger, resembling small cargo airplanes without motors.) It was a terrible night. Huge ore deposits in the mountains they passed seem to have disoriented the compass of one of the planes, guiding the pilot into a mountain edge.

The pilot of the other team's glider was an Australian, who found himself caught in an impossible dilemma in this disorienting northern hemisphere snowstorm at night: if he stayed with the towing Halifax when it was high, his wings and cable lines would ice up so heavily that he would crash. But if he released and flew low too early, swirling gales in the mountains would toss him away from any controlled path. The Australian's glider finally released, in heavy cloud, but something went wrong and it too came down in a heavy crash landing.

At each crash site there were a number of survivors, and in both cases a few of the troops—injecting themselves with morphine for their injuries; popping amphetamines to get through the snow—managed to reach local farmhouses for help. But all were soon arrested by German troops or local collaborators. Most were shot immediately; the others were tormented for a few weeks first.

. . .

Just a few years earlier, R. V. Jones had been a promising astronomy researcher at Balliol College, Oxford. Now, barely out of his twenties, he was director of intelligence on the air staff: faced with the sort of ethical dilemma that is an occasion for cleverness at an Oxford dinner; haunting in real life. Thirty Airborne specialists had been sent in, and every single one was dead. The factory hadn't even been reached.

"It fell to me," Jones remembered decades later, "to say whether or not a second raid should be called for. It came all the harder because I should be safe in London, whatever happened to the second raid, and this seemed a singularly unfitting qualification for sending another 30 men to their deaths. . . .

"I reasoned that we had already decided, before the tragedy of the first raid and therefore free from sentiment, that the heavy water plant must be destroyed; casualties must be expected in war, and so if we were right in asking for the first raid we were probably right in asking that it be repeated."

This time the Norwegians themselves took over. Six volunteers who'd made it to Britain were selected. One was an Oslo plumber, another had been an ordinary mechanic. Contemporary records suggest that the British had little confidence the Norwegians would succeed where dozens of crack Airborne troops had failed. Minimal attention, for example, was given to their possible escape afterward. But what else was there to do? As more heavy water continued to be shipped to Germany, work in Leipzig could progress; the Virus House unit in Berlin could catch up as well.

The six Norwegians were trained as well as possible, then sent to a luxurious safe house—S.O.E. Special

Training School Number 61—outside Cambridge for final preparations, and also to wait for the weather to clear. There were chatty English girlfriends, and the occasional dinner out in Cambridge. Then in February 1943 the meteorological reports improved; the house suddenly emptied.

After being parachuted in to Norway, they met up with an advance party of a few other Norwegians, which had waited in isolated huts all winter. Together, on cross-country skis, they reached Vemork a few weeks later, at about 9 P.M. on a Sunday night.

"Halfway down we sighted our objective for the first time, below us on the other side. . . . The colossus lay like a medieval castle, built in the most inaccessible place, protected by precipices and rivers."

It was the furthermost ripple of what had begun in Einstein's quiet thoughts: a handful of armed Norwegian men, panting in deep snow, staring at a lit fortress in the night. It was clear why the Germans had left only a small guard. The only way in was across the single suspension bridge, over an impassable stone gorge several hundred feet deep. It might be possible in a strong firefight to kill the guards on the bridge, despite their protected emplacements, but if that happened, the Germans would simply start killing the local townspeople. Both sides knew this. When a radio transmitter had been uncovered on Telavaag island the year before, every house and boat was burned, and all the women, and all the children—and of course all the men—who'd lived on the island were sent to concentration camps. Jones in London would probably not accept that again; the nine Norwegians looking down on the factory now definitely wouldn't. But this didn't mean they were going to go back. They had another way in.

From aerial reconnaissance photographs, highly

magnified in England, one of the team—Knut Haukelid—had noticed a clump of scrub plants a little further
along the gorge. "Where trees grow," he'd remarked, "a
man can make his way." One of their members had re-
connoitered the day before, to confirm this. They
started the climb down, cursing their heavy backpacks,
then crossed the river, which was ominously oozing water above its ice, and then cursed their backpacks even
more on the climb up to the factory. Since no one
wanted to disappoint the others, they all furtively
quickened their pace; the speed was soon exhausting.

Outside the factory perimeter they had to rest, sharing chocolate to get some strength. There was a loud
noise of turbines, for due to the orders from Leipzig and
Berlin, the factory worked on a twenty-four-hour schedule. What do nine highly armed men talk about? One
was teased for how he was trying to pick rations out
from between his teeth without the others noticing;
others spoke, more seriously now, about two young
married couples they'd met on the final night before
their skiing journey to Vemork. One of the parachuted
fighters had been at school with the young man in one
of the couples, but at first they'd been scared at coming
across armed strangers: they hadn't recognized him.
Then when they finally did, each side had realized it was
too dangerous to talk, even though the parachuted new-
comers were desperate to hear what ordinary life had
been like in Norway this past year. They'd had to spend
the night aware of the lamps on in the couples' cabin,
and the sight of smoke from their hearth fire; busying
themselves so they would have no thoughts of home;
just checking rifles and grenades and explosives, and
waxing their cross-country skis for this assault.

One of the men looked at his watch; the short rest
was over. They lifted their packs, and went to the gates.

There were advantages to having a big ex-plumber with them, for he now took out oversized wire-cutters and snapped right through the iron. They were inside.

It was the central moment. Heisenberg and the German army's Weapons Bureau had been constructing a "machine": a vast apparatus composed of uranium, and trained physicists, and engineers, and electricity supplies, and containment vessels, and neutron sources. Only when every part was in place could the mass from the center of uranium atoms be sucked out of existence, to be replaced by roaring energy in fast, unstoppable $E=mc^2$ explosions. The heavy water that controlled the flight of the triggering neutrons, slowing them down enough to "ignite" the uranium fuel, was the last part of this machine that had to be put in place. Germany's power—of troops and radar stations and local collaborators and SS inquisitors—had swatted down the British Airborne forces that had tried to obstruct the "machine" that would allow the power of $E=mc^2$ to emerge.

The nine Norwegian men were now all that were left. One group took up positions outside the guard barracks. Others watched the huge main doors to the factory. Blasting those open would have been possible, but again would have resulted in reprisals. An engineer who'd worked at the factory, though, had told the Resistance about a little-used cable duct that went in from the side. Two of the team, now loaded with all the explosives they'd carried, found it and crawled in.

The workers inside had no love for I. G. Farben, and were only too willing to let them go ahead. Within about ten minutes the charges were set. The workers were sent out, and the two men quickly followed.

At about 1 A.M., there was a slight thud; a brief flash at a few of the windows. The eighteen "cells" that separated out the heavy water were chest high and built of thick steel, looking a bit like overbuilt gas-fired boilers. No explosives that nine men could carry in their climb would totally destroy them. Instead, the Norwegians had set small plastic charges at the bottom of each one. The charges opened up holes, and also sent enough shrapnel flying out to cut exposed pipes.

The warm wind known as the *foehn* had started blowing, and the Norwegians could feel the snow starting to melt on their way back down the gorge. Searchlights came on as well as the air raid sirens, but this didn't matter. The terrain was rough enough to cover the men. As they climbed and then skied away, the heavy water gushed from the factory's drains, rejoining the mountain's streams.

$$E=mc^2$$

The raid bought time for the Allies, but even that would have been wasted if the wrong person had headed their project to build a bomb. At one point the Berkeley physicist Ernest Lawrence's name had been mooted, but his personnel skills made Heisenberg look considerate. America's own physics establishment had been so weak in the 1920s and 1930s that any bomb would have to be constructed, in large parts, by more highly skilled refugees from Europe. No one could have been worse to lead such a team than the broad-shouldered South Dakotan Lawrence.

In 1938, the Italian refugee Emilio Segrè had obtained a position at $300 per month in Lawrence's lab. It was a godsend for Segrè, who was Jewish, since if he and his young wife had to go back to Italy, there would be no possibility of working in a university anymore; there was also a good chance that they would be turned over to the Germans, and—as happened to many of their relatives—their children would quite possibly be killed. Segrè recalls what Lawrence did:

In July 1939, Lawrence, who by then must have realized my situation, asked me if I could return to Palermo. I answered by telling him the truth, and he immediately interjected: "But then why should I pay you $300 per month? From now on I will give you $116."

I was stunned, and even now, so many years afterward, I marvel . . . [that] he did not think for a second of the impression he conveyed.

The man who was appointed to the overall charge of the atomic bomb program, Leslie Groves, was somewhat better than Lawrence, at least in the sense that he wasn't prone to threatening his staff with imminent death. He also—like Lawrence—was effective in getting things built. He'd done a stint at MIT, then finished fourth in his class at West Point and had been largely responsible for getting the Pentagon building completed. Before the atomic bomb project was done, a vast reactor would need to be built, sited by a large river to take away the cooling water; factories thousands of feet long would have to be constructed, able to filter toxic uranium clouds. Groves got them all done, on time and under budget.

But Groves also carried a constant personal anger, of a sort more accepted in American public life at the time. He screamed, he threatened; he demeaned his assistants in public; his neck veins popped out with anger a lot. (The fact that he was now dealing with theoretical physicists of an intellectual level that dwarfed the accomplishments he'd been so proud of at West Point did not make him any easier to live with.)

When the secret Los Alamos, New Mexico, research center for the bomb project officially opened, in April 1943, Groves stood up to speak. A young member of the audience, Robert Wilson, later remembered: "He said he appeared not to believe in the eventual success of the project. He emphasized that if—or when—we failed, it could be he who would have to stand before a congressional committee to explain how money had been squandered. He could not have done worse at starting the conference on an upbeat note of enthusiasm."

Many possible projects have collapsed when administrators like this took over. A workable jet prototype, for example, had been up and running in Britain as early as 1941, but incompetent organization kept it from ever being deployed in sufficient numbers to help the RAF. Groves could motivate construction engineers who had to follow blueprints, yet he would almost certainly have failed at inspiring theoreticians who had to trust that they would succeed in unexplored intellectual terrain. But in the autumn of 1942—while Heisenberg was readying further work after his successful Leipzig tests— Groves made an appointment of genius. He selected the exquisitely oversensitive J. Robert Oppenheimer to be in day-to-day control of the scientists at Los Alamos.

It was a job that nearly destroyed Oppenheimer's health; by the time the first explosion took place, Oppenheimer, about six feet one, was down to about 116 pounds. In time his work on the Manhattan Project destroyed his career, making him so much of an outcast in the U.S. that he would have been jailed if he tried to read his own past classified papers. But he got the job done.

Oppenheimer's great strength, curiously enough, came from his underlying lack of confidence. It wasn't something most people could tell on the surface, of course. He had graduated from Harvard in only three years, with perfect grades; had studied at Rutherford's lab; taken his doctorate from Göttingen, and quickly after, still in his twenties, become one of America's top

J. Robert Oppenheimer

BULLETIN OF THE ATOMIC SCIENTISTS, AIP EMILIO SEGRÈ VISUAL ARCHIVES

theoretical physicists. He seemed effortlessly good at everything. He once asked a graduate student, Leo Nedelsky, to take over some of his lectures at Berkeley: "'It won't be any trouble,' said Oppenheimer, '. . . it's all in a book.'" Finding that the book was in Dutch, which he could not read, Nedelsky demurred. "'But it's such easy Dutch,' said Oppenheimer."

Yet it was all a fragile, frantic, uncertain ability. His whole family had been like that. His father had climbed up in the New York rag trade, then married a genteel woman who insisted her family do everything "properly": there were summerhouses, and servants, and classical music. At summer camp she saw to it that the other little boys were instructed to play with her Robert, and was surprised that he ended up being bullied, on one occasion being locked naked in the icehouse overnight. At Rutherford's lab he'd been so desperate at not being the top researcher that in a fit he'd tried to strangle his one friend. At Göttingen he'd had books hand-bound for himself, chided a graduate student couple for what he called their "peasant" ways in not being able to afford a baby-sitter—and then agonized over why people thought he was putting on airs.

As a result, Oppenheimer was superb at identifying weaknesses or inner doubts in others. When he lashed out at fellow researchers throughout his time as a Berkeley professor, he could unerringly select whatever area they felt weakest about, for he knew very well what it was to have an area to feel weak about. Even in his own physics he was aware of his own weaknesses, and felt a crushing sense of self-loathing at the way he regularly pulled back, ever so slightly, just when he might make a major breakthrough.

And then, at Los Alamos, he switched. The sarcasm was dropped, for the duration of the war. But the ability to detect other people's deepest fears or desires remained, and this meant that he became a superb leader of men.

He knew—instantly—that the young postgraduate physicists he needed in large numbers wouldn't pass up work at MIT's radar lab or other famous wartime projects to head to this unknown New Mexico site, simply on the basis of salaries, or offers of future jobs. They'd come only if they thought the top physicists in America were going there. Oppenheimer, accordingly, recruited the senior physicists first; the postgrads followed fast. He even got the authority-resistant genius Richard Feynman on his side. (Tell Feynman that something was a national emergency and his country needed him, and he'd give his mocking New York snort and tell you to get lost.) But Oppenheimer understood that Feynman was so hostile in large part out of furious anger: his young wife had tuberculosis, and in this era before antibiotics it was likely she would soon die. Oppenheimer obtained a rare-as-gold wartime train pass so she could come to New Mexico; he also arranged a place in a hospital close enough to Los Alamos so that Feynman could visit her regularly. In his later memoirs, Feynman joyously mocked every administrator he worked for—with the exception of the two years at Los Alamos, where he did everything Oppenheimer asked.

Oppenheimer's skills came to the fore in the hardest problem Los Alamos needed to solve. America was building two entirely different sorts of bombs. One team, led by Lawrence in Tennessee, took a blunt approach, and was simply trying to extract the most explosive component in natural uranium. When enough of that was accumulated, there'd be a bomb. The Tennessee factories followed the sort of straightforward engineering that Lawrence and other plain-talking Americans liked. Although there were exceptions, it was largely pushed by native-born Americans.

Another team, up in Washington State, was taking a more subtle approach. They were starting with ordinary uranium, and then hoping to transform that to an entirely new element, in a process of transmutation much like the one medieval alchemists and even Newton had struggled with in past centuries. The alchemists had wanted to turn lead into gold. The Washington State team, if they succeeded, would transform ordinary uranium into the wickedly powerful, new plutonium metal. Although again there were exceptions, this abstruse approach had been promoted more by the European refugees, educated in a more theoretical tradition.

The Pentagon liked Lawrence and the blunt Americans down in Tennessee, but it turned out that the foreigners' Washington project did best of all. Despite all of Lawrence's screaming and haranguing and threats, even after months of operation the Tennessee factories—giant factories, over a mile in total length; costing over a billion dollars (even in 1940s currency)—could barely sieve out enough purified uranium to stuff into a single envelope. No one was going to be able to make a bomb with that.

But although the Washington team did manage to create its promised plutonium, pretty soon the Los Alamos staff realized that no one could get it to ignite as a bomb. The problem wasn't that plutonium didn't explode. Rather, this new element exploded too *easily*. To make a simple uranium bomb—if the Tennessee team ever got enough purified uranium together—wouldn't be hard. If the amount that would make an explosion was 50 pounds, then you could make a 40-pound ball, and carve a hole in it, and then get a big gun, aim it at the hole, and fire—fast!—the remaining 10 pounds into it. The threshold would be reached so quickly, and the reaction would take place in such a small concentrated area, that much of the explosive U235 form of uranium would convert into energy before it blew itself apart.

The fragile and new plutonium was different. Fire two segments at each other and the plutonium would start exploding before the two halves completely clanged together. You wouldn't want to stand nearby when this started, of course, for there would be a gush of liquefied or gaseous plutonium where the reaction began. But that would be all. There would be almost no nuclear reaction: most of the raw plutonium, not transformed, would simply spatter away.

This is where Oppenheimer's insight and managerial gift came in. Forget about trying to clang two separate pieces of plutonium together. The way to get the plutonium fuel from Washington State to work, he realized, would be to start with a ball of plutonium that was fairly low density. That wouldn't explode. But then you'd wrap explosives around it, and set them off, all at *precisely* the same instant. Do it right, and the ball would crumple inward, so fast that the cascading sequence of $E=mc^2$ blasts that started spreading within would have enough time to accumulate before the plutonium flew apart.

The technique was called implosion, but the calculations were so hard—how do you make sure the plutonium ball doesn't crumple unevenly?—that there was a great deal of cynicism about whether it could work. (When Feynman first saw what the implosion theorists were trying, he pronounced simply: "It stinks!") Oppenheimer overcame that. He nurtured the first theorists who proposed implosion; he assembled the right explosives experts; as the project grew to a level that under anyone else's supervision it might have fallen apart in a mess of squabbling egos, he deftly manipulated the participants so that all the different groups involved worked together in parallel.

At one point he had the top U.S. explosives expert, and the top UK explosives expert, and the Hungarian John von Neumann—the quickest mathematician anyone had met, who would also help create the computer in his long career—and a host of other nationalities all working on it. He even had Feynman joining in! The one prima donna who might have destroyed the effort was the embarrassingly egocentric Hungarian physicist Edward Teller. Oppenheimer neatly led him away, and granted him his own office and work team, even amid the shortages of skilled manpower, to concentrate on his own prize ideas. Teller was vain enough—as Oppenheimer of course understood—that he simply took it as his due; in his pleasure he no longer bothered everyone else.

Paralleling the whole team was a purely British effort, which touched on these theoretical matters as well as practical isotope separation, at Chalk River, near Ottawa. Groves had been suspicious of this group, but Oppenheimer wanted all the help he could get.

Money didn't count. Everyone knew the level Germany was starting from. At one time, at Los Alamos, calculations suggested that a casing of solid gold might help bounce escaped neutrons back into an exploding bomb. (Its weight would also help keep the exploding plutonium bomb intact.) A little later, Charlotte Serber, who ran the library cum document storage room at Los Alamos, received a small package, about the size of a brown paper lunch bag.

"All that day Serber amused herself and the women who worked for her by saying to innocent would-be readers 'Please move these little packages to the next table for me.'

But they couldn't move the one that came from Fort Knox. Gold is denser than lead (that's why it was chosen), and the little 6-inch solid gold sphere inside weighed as much as an eighty-pound barbell.

But yet, despite the dozens of top researchers and the nearly unlimited funds, the plutonium problem still wasn't being solved. Was it possible, Oppenheimer and others worried, that no full bomb could be made this way? In that case, the best that might result would be an accumulation of radioactive plutonium. Maybe that was even what Heisenberg's heavy water reactor was being designed to cook up. Oppenheimer was informed, in a memorandum of August 21, 1943:

> It is possible . . . that [the Germans] will have a production, let us say, of two gadgets a month. This would place particularly Britain in an extremely

serious position but there would be hope for counter-
action from our side before the war is lost. . . .

One of the memo authors was Teller, who could be
discounted, but the other was Hans Bethe, an eminently
sensible man. He was the head of the theoretical division at Los Alamos, and he'd been a faculty member
with Geiger until 1933 in Tübingen. He had excellent
contacts with physicists who'd remained on the Continent. The "gadgets" Bethe and Teller had in mind were
full bombs, which were unlikely at this stage, but who
knew what else the Germans might build?

Even a few pounds of powdered radioactive metal
released over London could make parts of that city uninhabitable for years. There already were worrying reports of the advanced delivery weapons Germany was
working on, and one of Heisenberg's men was later seen
at Peenemünde, where the supersonic "vengeance"
weapon—the V-2 missile—was being built. Much simpler
jet drones were also being constructed—the V-1s—and if
those crashed highly radioactive warheads into Allied
troop emplacements, in the south of England before
D-Day or in France afterward, there could be casualties
of a level that had never before been seen.

The threats were taken so seriously that Eisenhower
accepted Geiger counters, and specialists trained in
their use, to be ready to go with his troops building up
in England for D Day. And then, at the very end of 1943,
when Oppenheimer was most lost in the plutonium implosion problem, Niels Bohr arrived at Los Alamos, after an escape from his institute in Copenhagen. Bohr
was the kindly elder statesman of physics. Over the
years everyone who counted, from Heisenberg to Oppenheimer to Meitner's nephew Robert Frisch, had
stayed at his institute and worked with him.

Now Bohr brought serious news. On December 6—after he'd fled—German military police had invaded his institute. They hadn't managed to steal the Nobel gold medals stored there, for George de Hevesy had dissolved them in a jar of strong acid, and left them—in liquid suspension—unobtrusively on a back shelf. But they had bullied their way around, arresting one of Bohr's colleagues who lived in the building. Most seriously, there were rumors that the institute's powerful cyclotron, an early form of particle accelerator, was going to be broken apart and sent back to Germany. Cyclotrons can make plutonium.

And then British military intelligence reported that, despite the sabotage and even a later Allied bombing raid, the factory at Vemork had been restarted. I. G. Farben engineers had been working frantically to repair it: replacement parts had been hurried in, and production now was higher than ever. In February 1944 the Norwegian Resistance reported that the entire heavy water stock was about to be sent back to Germany.

What to do? It was an excruciating moment, previewing the dilemmas the Allied physicists would face in the decision to use the bomb one year later. Another direct assault wasn't possible, for the Vemork factory was too heavily barricaded. The main train tracks out were heavily guarded as well—there were regular Army troops; SS detachments; auxiliary airfields that would be opened for spotter aircraft.

The sole weak spot for attacking the shipment back to Germany was where the train cars with the heavy water from Vemork had to be loaded onto a ferry to cross Lake Tinnsjö, on their way to the Norwegian coast. That was scheduled to take place in mid-February 1944.

If the train was sunk while it was on the ferry, no German divers could bring it up from the lake's depths. But Tinnsjö was also the main crossing to the rest of Norway for the factory workers at Vemork plus their families; it was also a popular tourist crossing. Ordinary families out for the day always were on the ferry.

Whom do you kill for a greater good?

Because of the equation—these powers $E=mc^2$ was offering—the physicists were demanding an awful moral trade-off, greater than anyone should be required to make. Knut Haukelid was one of the Norwegians who had remained behind after the factory raid, living rough on the Hardanger plateau, surviving massive manhunts. By now he was very experienced at the skills needed for sabotage: smuggling himself into a town; working out whom to trust; assembling and testing whatever explosives and timers would be needed. But that wasn't the issue. He had traveled this far, and lived this harshly, to save his countrymen. Now he would be killing them, drowning them in deep cold water.

Norway command to London:

REPORTS AS FOLLOWS: . . . DOUBT IF RESULT OF OPERATION IS WORTH REPRISALS STOP WE CANNOT DECIDE HOW IMPORTANT THE OPERATIONS ARE STOP PLEASE REPLY THIS EVENING IF POSSIBLE STOP

London to Norway command:

MATTER HAS BEEN CONSIDERED STOP IT IS THOUGHT VERY IMPORTANT THAT THE HEAVY WATER SHALL BE DESTROYED STOP HOPE IT CAN BE DONE WITHOUT TOO DISASTROUS RESULTS STOP SEND OUR BEST WISHES FOR SUCCESS IN THE WORK STOP GREETINGS

The best Haukelid could do was arrange with the Vemork transport engineer that the shipment would only come out on Sunday the 20th, when traffic would be light. (Trade union activity had always been strong in Vemork, and as a result the Resistance had high membership—and higher support—in the factory.) Late the Saturday night before, Haukelid arrived with two locals at the berthed ferry. They got on board safely, but when they were hunting for a spot below-decks to lay their explosives, the night watchman, a young Norwegian, found them. He knew one of Haukelid's companions, though, from a local sports club, and quickly nodded his agreement when they gave him their cover story: that Haukelid and the other man, Rolf Sörlie, had to hide from the Germans, and needed somewhere to store their packages. While the first two men stayed behind talking, Haukelid and Sörlie set the charges: right against the front hull, so the explosion would tip the boat forward, lifting the propeller uselessly up in the air, and would cause the tipped boat to fill with water and sink immediately. It was a half hour before Haukelid was done.

> When I left the watchman, I was not clear in my mind as to what I ought to do. . . . I remembered the fate of the two Norwegian guards at Vemork, who had been sent to a German concentration camp after the attack there. I did not want to hand over a Norwegian to the Germans. But if the watchman disappeared, there was danger of the Germans' suspicions being aroused next morning.
>
> I contented myself with shaking hands with the watchman and thanking him—which obviously puzzled him.

Lake Tinnsjö ferry
NORSK HYDRO

Everyone involved was in Haukelid's position. Alf Larsen, chief engineer at Vemork, had been at a dinner party earlier that evening, where a visiting violinist said that he'd be taking the boat the next day. Larsen had tried to say no, that he should stay longer in this beautiful region, the skiing was so excellent. But when the violinist waved that off, Larsen hadn't been able to insist. A contact at the factory had told him that his elderly mother, too, was planning to take the ferry.

The bomb went off at 10:45 A.M.; the boat was in 1,300 feet of water. The flatcars from the train broke loose in the sudden tilt, their doors bursting open. The factory worker's mother wasn't on board—her son hadn't let her out of the house—but the violinist was. There were fifty-three people on board. Most of the sturdy German guards managed to fight their way off the tumbling ship in time, but many of the women and children were pushed aside. Over a dozen passengers were caught inside.

A few of the barrels that had been only slightly purified bobbed on the top of the lake, and the passengers who'd managed to get off but hadn't made it into lifeboats—although the violinist did—grabbed on till a rescue boat came. But the barrels that contained the concentrated heavy water demonstrated, in slow-motion free fall, what they contained. Since the H_2O molecules are composed of a nucleus heavier than ordinary water, the barrels sank as if weighted, swirling around the ferry and its innocent trapped passengers down to the bottom.

One year and six months later, in August 1945, 50 pounds of purified Uranium 235, encased within 10,000 pounds of cordite, steel tamper, casing, and firing controls, was waiting on a heavy trolley, about to be loaded onto a B-29 on the island of Tinian, six hours' flying time from Japan. Oppenheimer was back in Los Alamos, monitoring this final operation.

If he were a simpler man, he might have been proud. The construction "machine" of researchers and factories and assembly units, which Heisenberg had abortively tried to put together in Germany, had here—on American shores, and under Oppenheimer's direction— finally been achieved. Rivers had been tapped to supply the processing plants and reactors; whole cities had been built to house tens of thousands of workers; a new element had been created through transmutation. It was an immense achievement.

Fermi's first neutron source, the one he'd used in Rome, based on Chadwick's design, could be held on the palm of one hand. The next device Fermi built, scraping together minor funds in New York in 1940, was about the size of a few large filing cabinets. By late 1942, with Oppenheimer overseeing the first substantial U.S. government funding, Fermi had built an enhanced device that filled much of a competition squash court, underneath the stands of the University of Chicago stadium. The final versions, constructed two years later, when atomic bomb funding was at full tilt, were the centerpieces of a 300,000-acre site in central Washington, near Hanford. With their supporting structures, they stretched taller than the entire Rome institute where Fermi had begun in 1934. Individuals who were aware of the full history could only stand in front of it in awe.

The plutonium problem had been solved, through mathematicians and explosives experts finding a shape for the ordinary explosives that would smoothly implode the plutonium ball. Regular supplies of the Washington site's output could now be machined for more bombs. The less successful Tennessee factories had also managed to produce a small amount of explosive, and it was Tennessee's total output—almost the complete amount of U235 the United States had—that was being loaded on Saipan.

Heisenberg's work had been blocked. Earlier in 1945, advancing Allied armies in Germany had found entire factories, some underground, with row upon row of completed jet-powered and even a few rocket-powered aircraft. But the Lake Tinnsjö sinking the previous year had guaranteed that only the barest amount of atomic construction could continue going forward. Even so, Heisenberg had tried to continue. Back in 1942, when funding had looked like it might slow down, he had eagerly explained the possible power of an atomic bomb to a conference of top Nazi administrators, in a quest to get funding back up. Now, even with the war near certain to be lost, he directed that the work be carried on from the small town of Hechingen, where he ended up lodging directly across the street from the home where Einstein's rich uncle had lived—the one who'd supported the family's business efforts, thereby giving Albert the subsidized years to prepare for university entrance.

The equipment lugged from Berlin and Leipzig had been ingeniously installed in a place observation planes wouldn't be able to find. It was put in a cave in an adjacent town, and the cave was in the side of a cliff, and on the top of the cliff was a church—and that was all you would see from the sky. Heisenberg had always been the one for grand gestures. When he'd first conceived of quantum mechanics, one night on a North Sea resort island at age twenty-four, he'd climbed the nearest dune peak and waited there till dawn, copying the Romantic characters from a Caspar David Friedrich painting. Now, in occasional excursions from the cave, he would climb to the highest point in the town, and go into the church, and there in his solitude play Bach with eloquent fury on the organ.

The atomic reactions had gone well beyond the old Leipzig work. By the end, the German researchers had reached about half the rate of nucleus splitting needed for a sustained chain reaction. Heisenberg knew he wouldn't get further. When a U.S. snatch squad did reach him in the Alps, even while Wehrmacht troops were still fighting in adjacent towns, he accepted surrender as if he'd been expecting it.

Heisenberg would be welcomed as a hero in Germany when he was finally released in 1946, while Oppenheimer, even before the war ended, knew his postwar life wouldn't be so simple. He had been a leftist in the late 1930s, and although a Berkeley physics professor might not suffer harm from that, once he was head of Los Alamos the FBI had dug up everything. Then he'd lied about some of the details, in his first interviews with military intelligence. Several important individuals wanted him out, but Groves was protecting him, so in revenge his enemies simply tormented him: much of his time as director his phone was tapped, his living quarters bugged, his past friends interrogated, and his trips shadowed. His wife had started drinking, a lot, and although he hadn't yet been attacked, he knew he was open to blackmail: the FBI had followed him on visits to San Francisco, where he'd spent nights with a girlfriend he'd been close to in the past.

More important, he knew what had happened on Lake Tinnsjö; he knew what was in store in the Pacific. It's common today to state that the atomic bombing of Japan was obviously justified, on the grounds that the alternative would have been an invasion that had to be much worse. But at the time it was not so clear. The bulk of Japan's army was no threat to American forces: it was sequestered up in China, with American submarines keeping it from crossing to the home islands, and the great weight of Russia's army looming above, able to destroy it once a sufficient buildup had occurred. Japan's industry had largely been burned out. Early in 1945, U.S. strategic bombers had been assigned the task of destroying thirty to sixty large and small cities. By August, they had burned out fifty-eight of them.

Douglas MacArthur, who had run much of the Pacific campaign, didn't expect an invasion would be needed; Admiral Leahy, chairman of the Joint Chiefs of Staff, was later adamant that there had been no need for an atomic bomb; Curtis LeMay, the head of the strategic bombing force, agreed. Even Eisenhower, who'd had no qualms about killing thousands of opponents when it was necessary to safeguard his troops, was strongly hostile to it, as he explained at the time to Henry Stimson, the elderly secretary of war: "I told him I was against it on two counts. First, the Japanese were ready to surrender and it wasn't necessary to hit them with that awful thing. Second, I hated to see our country be the first to use such a weapon. Well . . . the old gentleman got furious. . . ."

The feeling it might not be needed was so strong that there was talk about having demonstrations first, or at least adjusting the phrasing in the surrender demands to make clear that the emperor could remain in place. Oppenheimer had been at many of those meetings: listening intently, arguing—often in a slightly hedged way—for use if it was needed, but supporting the clause about safeguarding the emperor.

These arguments didn't take. Truman's most forceful adviser was Jimmy Byrnes, a man of Lyman J. Briggs's generation, but far less mild in temperament. The ethos Byrnes had been brought up with was that when you fought, you fought with everything you had. He'd been raised in South Carolina in the 1880s, with no father and not a great deal of schooling. Visitors to his state during earlier times reported their amazement that it was rare on a jury to find twelve men who had all their eyes and ears: South Carolina still had the ethos of a frontier society, and gouging, biting, and knife slashes were the way fights were settled. It was Byrnes who ensured that the clause protecting the emperor—which might mollify Japanese opponents of a settlement—was taken out. There would be no nonsense either about just waiting for the submarine blockade to tighten, or getting advancing Russians to do the dirty work.

Notes from the Presidential "Interim Committee," June 1, 1945:

> *Mr. Byrnes recommended,* and the Committee *agreed,* that . . . the bomb should be used against Japan as soon as possible; that it be used on a war plant surrounded by workers' homes; and that it be used without prior warning."

Part of Oppenheimer accepted that; part of him—especially when away from Washington—was unsure. But did it matter? He'd helped bring out these powers but now was the least part of it. Oppenheimer's superior, Leslie Groves, was General Groves. Los Alamos was a project of the United States Army. The army built weapons to use them.

The atomic bomb was going to be loaded onto that airplane.

13

E=mc²

Whistling, spinning, the bomb ("an elongated trash can with fins") had taken forty-three seconds to fall from the B-29 that released it. There were small holes around its midpoint where wires had been tugged out of it as it dropped away: this had started the clock switches of its first arming system. More small holes had been drilled farther back on its dark steel casing, in New Mexico, and those took in samples of air as the free fall continued. When it had tumbled to 7,000 feet above the ground, a barometric switch was turned, priming the second arming system.

From the ground the B-29 was just visible as a silvery outline, but the bomb—a bare ten feet long, two and a half feet wide—would have been too small a speck to see. Weak radio signals were being pumped down from the bomb to the Shina Hospital directly below. Some of those radio signals were absorbed in the hospital's walls, but most were bounced back skyward. Sticking out of the bomb's back, near the spinning fins, were a number of whiplike thin radio antennae. Those collected the returning radio signals, and used the time lag each took to return as a way of measuring the height remaining to the ground.

At 1,900 feet the last rebounded radio signal arrived. John von Neumann and others had calculated that a bomb exploding much higher would dissipate much of its heat in the open air; exploding much lower, it would dig a huge crater in the ground. At just under 2,000 feet the height would be ideal.

An electric impulse lit cordite sacs, producing a conventional artillery blast. A small part of the total purified uranium was now pushed forward down a gun barrel that was actually inside the bomb. In the early planning this gun had been a very heavy device, being simply a copy of large U.S. Navy weapons. Only after several months had one of Oppenheimer's men realized that navy guns were so heavy because they had to survive the recoil of shot after shot. Here, of course, it wouldn't matter: this gun was only going to be fired once. Instead of weighing 5,000 pounds it was machined to weigh barely a fifth of that.

The first uranium segment traveled about four feet within the thinned gun barrel, and then it impacted the remaining bulk of the uranium. Nowhere on Earth had a ball of several dozen pounds of such purified uranium ever been accumulated. There were a number of stray neutrons loose inside it, and although the uranium atoms were densely protected by their outer flurries of electrons, the escaped neutrons, having no electrical charge, weren't affected by the electrons. They flew through the outer electron barrier—as we saw, like a probe skimming past the planets down toward our sun—and while many of them flew straight through out the other side, a few were on a collision course for the speck of a nucleus far down at the center.

That nucleus normally blocked outside particles

from entering, for it was seething with positively charged protons. But since neutrons have no electric charge they're invisible to the protons as well. The arriving neutrons pushed in to the nucleus, overbalancing it; making it jostle and wobble.

The uranium atoms mined on Earth were each over 4.5 billion years old. Only a very powerful force, before the Earth was formed, had been able to squeeze their electrically crackling protons together. Once that uranium had been formed, the strong nuclear force had acted, gluelike, to hold these protons in place over all that long span: while the Earth cooled, and continents formed; as America separated from Europe, and the North Atlantic Ocean slowly filled; as volcanic bursts widened on the other side of the globe, forming what would become Japan. A single extra neutron unbalanced that stability now.

Once the wobbling in the nucleus was enough to break the strong force glue, then the ordinary electricity of the protons was available to force them apart. A single nucleus doesn't weigh much, and the fragmentary section of one weighs even less. Its speeding impact into the other parts of the uranium didn't heat it up much. But the density of uranium was enough that a chain reaction started, and soon there weren't just two speeding fragments of uranium nuclei, there were four, then eight, then sixteen, and so on. Mass was "disappearing" within the atoms, and coming out as the energy of speeding nuclei fragments. $E=mc^2$ was now under way.

The entire sequence of multiplying releases was finished in barely a few millionths of a second. The bomb was still suspended in the humid morning air with a faint layering of condensation on its outer surface, for it had been up in the cold air of 31,000 feet just forty-three seconds earlier, and now, 1,900 feet over the hospital, it was a balmy 80°F. The bomb fell downward just an additional fraction of an inch in the time of most of the reaction; from the outside there would only be the first odd bucklings of its steel surface to suggest what was going on inside.

The chain reaction went through eighty "generations" of doubling before it ended. By the last few of those, the segments of broken uranium nuclei were so abundant, and moving so fast, that they started heating up the metal around them. The last few doublings were the crucial ones. Imagine you have a pond in your garden, with a lily plant floating on it that doubles in size every day. In eighty days the lily entirely covers the pond. On which day is half the pond still uncovered, open to the sun and outside air? It's the seventy-ninth day.

From this point on, all the action of the $E=mc^2$ reaction was over. No more mass was "disappearing"; no more fresh energy appeared. The energy in the movement of those nuclei was simply being transformed to heat energy—just as rubbing your hands together will make your palms warm up. But the uranium fragments were rubbing against resting metal at immense speed, due to the multiplication by c^2. They soon were traveling at a substantial fraction of the speed of light.

The rubbing and battering made the metals inside the bomb begin to warm. They had started at near body temperature—98.6°F or 37°C—and then they reached water's boiling temperature—212°F or 100°C—and then that of lead—560°C. But the generations of chain reaction doubling had gone on, as yet more uranium atoms had been splitting, so it reached 5,000°C (the surface of the sun) and then several million degrees (the temperature of the center of the sun) and then it kept on rising. For a brief period, in the center of the suspended bomb, conditions similar to those in the early moments of the creation of the universe were produced.

The heat moves out. It goes through the steel tamping around the uranium, and just as easily through what had been the several-thousand-pound massive casing of the bomb, but then it pauses. Entities as hot as that explosion have energy that must be released. It starts pushing X rays out of itself, a very large number of them, some of them angling up, and some to the side, and the rest in a wide stretching arc toward the ground.

The explosion is hovering; the fragments are trying to cool themselves off. They remain that way, pouring out a large part of their energy. Then, after 1/10,000th of a second, when the X ray spraying is over, the heat ball resumes its outward spread.

Only now does the central eruption become visible. Ordinary light photons could not push through the X ray sprays; only the glows on the outside of the sprays would have been seen. When the full flash appears, it's as if a rip in the sky has opened. An object resembling one of the giant suns from a distant part of our galaxy now appears. It fills several hundred times more of the sky than Earth's ordinary sun.

The unearthly object burns at full power for about one-half of a second, then begins to fade away, taking two or three seconds to empty itself out. This "emptying" is accomplished, in large part, by spraying heat energy outward. Fires begin, seemingly instantaneously; skin explodes off, hanging in great sheets from the bodies of everyone below. The first of the tens of thousands of deaths in Hiroshima begin.

At least a third of the energy from the chain reactions comes out in this flash. The rest now follows soon behind. The strange object's heat pushes on ordinary air, accelerating it to speeds that have never occurred

An atomic bomb exploding in the very first milliseconds (top)
and the ground being churned up as it expands, prior to the bomb
producing a mushroom cloud (center; bottom).

here before, unless at some time in the distant past a large meteor or comet arrived. It travels several times faster than any hurricane could achieve—so fast, in fact, that it's silent, for it outruns any sound its immense force might make. After it there's a second air pulse, a little slower; after that the atmosphere sloshes backward, to fill up the gap pushed out. This briefly lowers the air density to virtually zero. Far enough from the blast, life-forms that have survived will now begin to explode outward, having been exposed—briefly—to the vacuum of outer space.

A small amount of the heat that was produced can't move forward at all. It remains behind, hovering quite close to where the fuses and antennae and cordite had been. In a few seconds it begins to rise. It swells as it goes, and at sufficient height it spreads out.

And when that great mushroom cloud appeared, $E=mc^2$'s first work on planet Earth was done.

PART 5

Till the End of Time

14

$$E=mc^2$$

The flash of light from the explosion over Hiroshima in 1945 reached the orbit of the moon. Some of it bounced back to Earth; much of the rest continued onward, traveling all the way to the sun, and then indefinitely beyond. The glare would have been viewable from Jupiter.

In the perspective of the galaxy, it was the most insignificant flicker.

Our sun, alone, explodes the equivalent of many million such bombs every second. For $E=mc^2$ does not apply just on Earth. All the scrambling commandos and anxious scientists and cold-eyed bureaucrats: all that is but a drop, the slightest added whisper, in the enormous powerful onrushing of the equation.

Einstein and other physicists had long recognized this; it was just a quirk that the accelerated technology and pressures of wartime had led to the equation's first applications being focused on weaponry. In this section of the book we switch to those wider views; lifting away from earthly technology, and showing how the equation's sway extends throughout the universe: controlling everything from how the first stars ignited, to how life will end.

. . .

Ever since the discovery of radioactivity in the 1890s, researchers had suspected that uranium or a similar fuel might be operating in the broader universe, and in particular, in our sun to keep it burning. Something that powerful was needed, because Darwin's insights as well as findings in geology had shown that Earth must have been in existence—and warmed by the sun—for billions of years. Coal or other conventional fuels would not be strong enough to do that.

Unfortunately, though, astronomers couldn't find any signs of uranium in the sun. Every element gives off a distinctive visual signal, and the optical device called the spectroscope (for it breaks apart the "spectrum") allows them to be identified. But point a spectroscope at the sun, and the signals are clear: there is no uranium or thorium or other known radioactively glowing element up there.

What did seem to leap out, in readings from distant stars as well as our own sun, was that there was always iron inside them: lots and lots of metallic bulky iron. By the time Einstein was finally able to leave the patent office, in 1909, the best evidence was that the sun was about 66 percent pure iron.

This was a disheartening result. Uranium could pour out energy in accord with $E=mc^2$, because the uranium nucleus is so large and overstuffed that it barely holds together. Iron is different. Its nucleus is one of the most perfect and most stable imaginable. A sphere made of iron—even if it was molten or gaseous or ionized iron—could not pour out heat for thousands of millions of years.

Suddenly the vision of using $E=mc^2$ and related equations to explain the whole universe was blocked.

Astronomers could just look past the top of the atmosphere, to the great spaces and waiting suns beyond us, and wonder.

The individual who broke that barrier—letting $E=mc^2$ slip the surly bonds of Earth—was a young Englishwoman named Cecilia Payne, who loved seeing how far her mind could take her. Unfortunately, the first teachers she found at Cambridge when she entered in 1919 had no interest in such explorations. She switched majors, and then switched again, which led to her reading up on astronomy, and when Payne decided on *anything*, the effects were impressive. She terrified the night assistant at the university's telescope her first night there, after she'd been reading for only a few days. He "fled down the stairs," she recalled, "gasping: 'There's a woman out there asking questions.'" But she wasn't put off, and a few weeks later she described another such incident: "I bicycled up to the Solar Physics Observatory with a question in my mind. I found a young man, his fair hair tumbling over his eyes, sitting astride the roof of one of the buildings, repairing it. 'I have come to ask,' I shouted up at him, 'why the Stark effect is not observed in stellar spectra.'"

This time her subject did not flee. He was an astronomer himself, Edward Milne, and they became friends. Payne tried to pull her arts student friends into her astronomical excitements, and even though they might not have understood much of what she was saying, she was the sort of person others like being around. Her rooms at Newnham College were almost always crowded. A friend wrote: ". . . when safely lying on her back on the floor (she despises armchairs), she will talk of all things under the sun, from ethics to a new theory of making cocoa."

Cecilia Payne
THE PRINCIPAL AND FELLOWS OF NEWNHAM COLLEGE, CAMBRIDGE, ENGLAND

Rutherford was teaching at Cambridge by then, but didn't know what to do with Payne. With men he was bluff and friendly, but with women he was bluff and pretty much a thug. He was cruel to her at lectures, trying to get all the male students to laugh at this one female in their midst. It didn't stop her from going—she could hold her own with his best students in tutorials—but even forty years later, retired from her professorship at Harvard, she remembered the rows of braying young men, nervously trying to do what their teacher expected of them.

But Arthur Eddington, a quiet Quaker, was also at the university, and he was happy to take her on as a tutorial student. Although his reserve never lifted—tea with students was generally in the presence of his elderly unmarried sister—the twenty-year-old Payne picked up Eddington's barely stated awe at the potential power of pure thought.

He liked to show how creatures who lived on a planet entirely shrouded in cloud would be able to deduce the main features of the unseen universe above them. There would have to be glowing spheres out there, he imagined them reasoning, for the original gas clouds floating in space would gradually form dense enough clumps to start nuclear reactions inside and light up—they would become suns. These glowing spheres would be dense enough to pull planets swinging around them. If the beings on the mythical planet ever did find that a sudden wind had blown an opening in their clouds, when they looked up they'd see a universe of glowing stars, with circling planets, just as they'd expected.

It was exhilarating to think that someone on Earth might solve the problem of how to deal with all the iron in the sun, and so be able actually to work out Eddington's vision. When Eddington first assigned Payne a problem on stellar interiors, which might at least start to achieve this, "the problem haunted me day and night. I recall a vivid dream that I was at the center of [the giant star] Betelgeuse, and that, as seen from there, the solution was perfectly plain; but it did not seem so in the light of day."

But even with this kind man's backing, a woman couldn't do graduate work in this field in England, so she went to Harvard, and there blossomed even more. She switched from her heavy woolen clothing to the lighter fashions of 1920s America; she found a thesis adviser, Harlow Shapley, an up-and-coming astrophysicist; she loved the liberty she found in the student dorms, and the fresh topics in the university seminars. She was bursting with enthusiasm.

And that could have blocked everything. Raw enthusiasm is dangerous for young researchers. If you're excited by a new field—keen to join in with what your professors and fellow students are doing—that usually means you'll be trying to fit in with their approaches.

But students whose work stands out usually have had some reason to avoid this, and keep a critical distance. Einstein didn't especially respect his Zurich professors: most, he thought, were drudges, who never questioned the foundations of their teaching. Faraday couldn't be content with explanations that left out the inner feelings of his religion; Lavoisier was offended at the vague, inexact chemistry handed down by his predecessors. For Payne, some of her needed distance came from getting to know her fun fellow Ivy League students a little better. Shortly after arriving: "I expressed to a friend that I liked one of the other girls in the House where I lived at Radcliffe College. She was shocked: 'But she's a Jew!' was her comment. This frankly puzzled me. . . . I found the same attitude towards those of African descent."

She also got a glimpse of what was going on in the back rooms at the Observatory. In 1923, the word *computer* did not mean an electrical machine. It meant people whose sole job was to compute. At Harvard, it was applied to ranks of slump-shouldered spinsters in those back rooms. A few of them had once had first-rate scientific talent ("I always wanted to learn the calculus," one said, "but [the director] did not wish it"), yet that was usually long since crushed out of them, as they were kept busy measuring star locations, or cataloging volumes of previous results. If they got married they could be fired; if they complained of their low salaries, they would be fired as well.

Lise Meitner had had her problems in getting started in research in Berlin, but there was nothing like this desolate, life-crushing sexism. A few of the Harvard "computers," in several decades of bent-back work, succeeded in measuring over 100,000 spectral lines. But what it meant, or how it fit in with the latest developments in physics, was almost always not for them to understand.

Payne was not going to be pushed into their ranks. Spectroscope readings can be ambiguous where they overlap. Payne began to wonder how much the way her professors broke them apart depended on what they already had in mind. For example, let the reader note the following letters very well, and then try to read them:

```
n o t e
v e r y
o n e w
i l l g
e t i t
```

It's not easy. But if you start reading it instead as "Not everyone . . ." then it leaps out. What Cecilia Payne decided on, there in 1920s Boston, was a Ph.D. project that would let her confirm and further develop a new theory about how to build up spectroscope interpretations. Her work was more complicated than our example above, for spectroscope lines from the sun will always include fragments of several elements; there are distortions from the great temperature as well.

An analogy can show what Payne did. If astronomers are convinced there's going to be lots of iron in the sun (which seemed fair for there was so much iron on Earth and in asteroids), there'd be only one way to read an ambiguous string of lines from a spectroscope. If they came out, for example, as:

```
t h e y s a i d i r o n a g a i e n
```

you'd parse it to read:

t h e y s a i d **I r o n** a g a i e n

and there'd be no need to worry too much about the odd spelling of *agaien*. The extra *e* could be a fault in the spectroscope, or some odd reaction on the sun, or just a fragment that was slipped in from some other element. There's always something that doesn't fit. But Payne kept an open mind. What if it was really trying to communicate:

t **H** e y s a i **d i r o** n a **g** a i **e** n

She went through the spectroscope lines over and over again, checking for these ambiguities. Everyone had boosted the lines one way, to make it read as if they were for iron. But it wasn't too much of a stretch to boost them differently, so that they read hydrogen, not iron.

Even before Payne finished her doctorate, her results began to spread in gossip among astrophysicists. While the old explanation of the spectroscope data had been that the sun was two-thirds iron or more, this young woman's interpretation was that it was over 90 percent *hydrogen,* with most of the rest being the nearly as light-weight helium. If she was right, it would change what was understood about how stars burn. Iron is so stable that no one could imagine it transformed through $E=mc^2$ to generate heat in our Sun. But who knew what hydrogen might do?

The old guard knew. Hydrogen would do nothing. It wasn't there, it couldn't be there; their careers—all their detailed calculations, and the power and patron-age that stemmed from it—depended on iron being what was in the sun. After all, hadn't this female only picked up the spectroscopic lines from the sun's outer atmosphere, rather than its deep interior? Maybe her readings were simply confused by the temperature shifts or chemical mixes there. Her thesis adviser declared her wrong, and then *his* old thesis adviser, the imperious Henry Norris Russell, declared her wrong, and against him there was very little recourse. Russell was an exceptionally pompous man, who would never accept he could be wrong—and he also controlled most grants and job appointments in astronomy on the East Coast.

For a while Payne tried to fight it anyway: repeating her evidence; showing the way her hydrogen interpretation was just as plausible in the spectroscope lines as the iron ones; even more, the way new insights—the latest in European theoretical physics—were suggesting a way hydrogen really could power the sun. It didn't matter. She even tried reaching out to Eddington, but he withdrew, possibly out of conviction, possibly out of caution before Russell—or possibly just from a middle-aged bachelor's fear of a young woman turning to him with emotion. Her friend from her student days at the Cambridge Solar Physics Observatory, the young fair-haired Edward Milne, was by now an established astronomer, and did try to help, but he didn't have enough power. Letters were exchanged between Payne and Russell, but if she wanted to get her research accepted she'd have to recant. In her own published thesis she had to insert the humiliating line: "The enormous abundance [of hydrogen] . . . is almost certainly not real."

A few years later, though, and the full power of Payne's work became clear, for independent research by other teams backed her spectroscope reinterpretations. She was vindicated, and her professors were shown to have been wrong.

. . .

Although Payne's teachers never really apologized, and tried to hold down her career as long as they could, the way was now open to applying $E=mc^2$ to explain the fires of the sun. She had shown that the right fuel was floating up in space; that the sun and all the stars we see actually are great $E=mc^2$ pumping stations. They seem to squeeze hydrogen mass entirely out of existence. But in fact they're simply squeezing it along the equals sign of the equation, so that what had appeared as mass, now bursts into the form of billowing, explosive energy. Several researchers made starts on the details, but the main work was done by Hans Bethe, the same man who later co-wrote that 1943 memo to Oppenheimer about the ongoing German threat.

Down on Earth, the many hydrogen atoms in our atmosphere just fly past each other. Even if crushed under a mountain of rock, they won't really stick to each other. But trapped near the center of the sun, under thousands of miles of weighty substance overhead, hydrogen nuclei can be squeezed close enough together that they will, in time, join together, to become the element helium.

If this were all that happened, it wouldn't be very important. But each time four nuclei of hydrogen get squeezed together, Bethe and the others now showed that they follow the potent, subatomic arithmetic of the sort Meitner and her nephew Frisch worked on that afternoon in the Swedish snow. The mass of the four hydrogen nuclei can be written as 1+1+1+1. But when they join together as helium, their sum is not equal to 4! Measure a helium nucleus very carefully, and it's about 0.7 percent less, or just 3.993 units of weight. That missing 0.7 percent comes out as roaring energy.

It seems like an insignificant fraction, but the sun is many thousands of times the size of Earth, and the hydrogen in this tremendous volume is available as fuel. The bomb over Japan had destroyed an entire city, simply from sucking several ounces of uranium out of existence, and transforming it into glowing energy. The reason the sun is so much more powerful is that it pumps 4 million *tons* of hydrogen into pure energy each second. One could see our sun's explosions clearly from the star Alpha Centauri, separated from us by 24 trillion miles of space; and from unimagined planets around stars far along the spiral arm of our galaxy as well.

The sun did that much pumping yesterday when you woke up—4 million tons of hydrogen "squeezed" along Einstein's 1905 equation from the mass side to the energy side, getting multiplied by the huge figure of c^2— and it was pouring out that much energy at dawn over Paris five centuries ago, and when Mohammed first took refuge in Medina, and when the Han Dynasty was established in China. Energy from millions of disappearing tons was roaring overhead each second when the dinosaurs lived: Earth has been nurtured, and warmed, and protected, by this same raging fire as long as it has been in orbit.

15

$E=mc^2$

Cecilia Payne's work had helped show that our sun and all the other stars in the heavens are great $E=mc^2$ pumping stations. But on its own, hydrogen-burning could easily have led to a sterile, dead universe. Early in the universe's history, there would have been great blazings as the hydrogen stars created their helium. But the original hydrogen fuel would eventually have used itself up, and the fires explained by $E=mc^2$ would have gradually died down, leaving only giant floating ash heaps of used helium. Nothing else would have ever been created.

To create the universe we do know, there had to be some device for building the carbon and oxygen and silicon and all the other elements that planets and life depend on. These elements are larger and more complex than what a simple hydrogen-to-helium combustion machine could ever produce.

Payne had been independent enough to challenge the consensus that stars were made of iron, and this had allowed the first stage of insight: showing that there actually was enough hydrogen in the stars above our atmosphere to allow the energy-spraying sequence of

$1 + 1 + 1 + 1 =$ not quite 4.00 to occur, thereby sustaining their fires. But with the production of helium, it stopped. Who would be cockily independent enough to go further, and show how $E=mc^2$ could operate to create the ordinary elements of our planet and daily life as well?

In 1923, when Payne arrived at Harvard, a seven-year-old Yorkshire lad was found by his local truant officer to have been spending most of the past year at the local cinema. Even though young Fred Hoyle explained most forcefully that it had been good for his education—he'd taught himself to read by following the subtitles—he was forced, against his will, back to school. It would be this young boy's work that ultimately solved the next major step in how the sun burns.

About one year after Hoyle returned to school, his class was assigned to collect wildflowers. Back in the classroom the teacher read out the list of flowers, describing one as having five petals. Hoyle examined the sample he'd collected, now in his hand. It had six petals. This was curious. If it had been a petal less than described, that would have been understandable, he might have torn one off in carrying it. But how could there be more? He was puzzling over it, and vaguely heard a strident voice, and then: "The blow was delivered flat-handed across the ear," he wrote, ". . . the one in which I was to become deaf in later life. Since, moreover, I wasn't expecting it at all, I had no opportunity to flinch by the half inch or so that would have reduced the impulsive pressure on my drum and middle ear."

It took a few minutes for Hoyle to recover, but then he left the school, and back at home explained to his mother what had happened: "I pointed out, I'd given the school system a tryout over three years, and, if you

Fred Hoyle

didn't know something was no good after three years, what did you know?"

His mother pretty much agreed, and so did his father, who had survived two years as a machine gunner on the Western Front by disobeying the less than brilliant orders from his upper-class officers to test-fire his guns at ten-minute intervals (which would have given his squad's exact location to German assault teams). Fred Hoyle got yet another year off. "Each morning, I ate breakfast and started off from home, just as if I were going to school. But it was to the factories and workshops of Bingley that I went. There were mills with clacking and thundering looms. There were blacksmiths and carpenters. . . . Everybody seemed amused to answer my questions."

In time he was railroaded back to school, where a few kind teachers saw his talent, and helped with scholarships. He ended up studying mathematics and then astrophysics at the University of Cambridge, and he did so well that the intensely private Paul Dirac took him on as a student, which was unheard of, and he even had tea with Payne's old supervisor Eddington—though as there were rumors of some sort of intellectual "disgrace" she had run into at Harvard, Payne's name was now barely mentioned. (History had been rewritten: Henry Norris Russell and the others now implied they'd "always" known that plenty of hydrogen was available in the sun.)

The problem of how stars manage to use helium as a further fuel in the giant $E=mc^2$ pumps, however, hadn't gone much further than where Payne's work and the direct follow-ups had left it in the 1920s. The heat of over 10-million-degrees at the center of our sun was able, barely, to squeeze the positive charges of four hydrogen nuclei together to make helium. To squeeze together those helium nuclei in a burning process to get larger elements, you'd need to get higher temperatures. But the universe was well surveyed.

Where could you find something hotter than the center of a star?

Hoyle's habit of putting things together in his own way now came to the fore. At the start of World War II he was sent to a radar research group, and in December 1944, after an information-sharing mission to the United States, he ended up waiting in Montreal for a rare flight back across the Atlantic.

He wandered around the city and beyond, and also picked up information about the British research group at Chalk River (about 100 miles from Ottawa). Although nobody told him anything official about the Manhattan Project, from the names he heard there—including several whose work he'd known at Cambridge before the war—he gradually deduced the basic stages of the top-secret project still going on at Los Alamos.

The easiest way to accumulate the raw material for a bomb, he already knew from reading accounts published before the war, was by cooking up plutonium in a reactor. He also knew that Britain had not tried building reactors. That meant, he concluded, that the specialists must have found some unsuspected problem with the plutonium route; probably with getting the ignition to operate fast enough. Now, though, seeing the specialists in Canada, including experts in the mathematics of explosions, he realized it must have been overcome.

Oppenheimer and Groves had barbed wire and armed guards and layers of security officers around the plutonium detonation group at Los Alamos. But that was no protection against a man who'd managed to outwit the stern educational establishment of village Yorkshire. By the time he was finally assigned a seat on a flight back, Hoyle had outlined what Oppenheimer's hundreds of specialists had proven. A substance such as plutonium that won't fully explode on its own will certainly crash apart its own atoms if it's squeezed inward abruptly enough. Implosion raises the pressure and temperature enough to do that.

Everyone in the bomb projects had thought of implosion as intensely localized; suitable only for plutonium spheres a few inches across. But why did it have to stay so small? Implosion was a powerful technique on Earth. Hoyle was used to following his thoughts anywhere. Why couldn't it apply in the stars?

If a star ever imploded, it too would get hotter. Instead of being below 20 million degrees, its center could reach—as Hoyle quickly computed—closer to 100 million degrees. That would be enough to squeeze even the larger nuclei of more massive elements together. Helium could be squeezed to create carbon. If the implosion went further, the star would get even hotter, and then heavier nuclei would be created: oxygen, silicon, sulfur, and the rest.

It all depended on a star's actually undergoing this inner collapse, but Hoyle realized there was a plausible reason this should happen. When a star was still at the relatively cool 20 million degrees, and capable only of burning hydrogen, the helium that was produced would build up like ash in a fireplace. When all the hydrogen was used up, that ash wouldn't be able to burn. The upper reaches of the star would no longer be pushed outward by the fires within. They would come crashing inward—just as in the Los Alamos bomb.

When a star implodes inward, that would raise its temperature to the 100 million degrees that is enough to ignite the helium ash. When *that* helium is used up, a further ash accumulates and the next stage occurs. The carbon can't burn at 100 million degrees, so now a further level of the star crashes down. The temperature gets higher, and the cycles go on. It's as if a multifloor building were slowly collapsing, as the struts holding up one floor after another suddenly buckle and break. $E=mc^2$ is central, for each level of burning—first the hydrogen, then the helium, then the carbon—gets its power from the conversion of mass into energy.

There were more details to come next, many contributed by Hoyle himself, but the idea taken from the atomic bomb had been central in solving the problem. Hoyle had simply switched the implosion process from the few pounds of plutonium laboriously collected on Earth, to a sphere of ultraboiling gas—a star—hundreds of thousands of miles wide, at immense distances away in space. He'd seen how stars can cook up the elements of life. When the larger of these stars used up their last possible fuel, it was also clear they'd have to break apart. Everything they had made would then pour out.

. . .

We tend to think of our planet as old, but when it was newly formed the heavens were already ancient; full of millions of these exploded giants. Their eruptions flung out silicon, and iron and even oxygen, to make the substance of Earth.

A large number of unstable elements such as uranium and thorium were created in the ancient stars' explosions as well, and when these elements floated over, becoming incorporated in the deep body of Earth, their continued explosions shot fragments of their nuclei at high velocity into the surrounding rock.

Along with the initial heat left over from the impacts of Earth's creation, the radioactive blastings from the uranium and similar heavy elements have kept our planet's depths from cooling. Their repeated multitudes of $E=mc^2$ bursts helped produce enough churning heat underneath the surface to make the thin continents on top roll forward, so shaping the surface of Earth.

In some places, sections of the thin crust were pushed crumpling into each other, producing the lifted ripples we call the Alps, Himalayas, or Andes. In other places, the churning heat pulled open gaps that we know by such names as San Francisco Bay, the Red Sea, and the Atlantic. These made excellent collection basins for the hydrogen that had also landed, and as that combined with oxygen, the result was oceans of sloshing water. Iron deep inside the planet sloshed in its own more stately fashion, driven by the daily spinning of the whole globe around its axis. That sent up invisibly streaking magnetic lines, of exactly the sort Michael Faraday described, and reproduced, in the basement of London's Royal Institution 4 billion years later. The result was an invisible network of magnetic force lines, far overhead, helping shield the self-assembling carbon molecules on the surface from some of the worst of the spraying radiation from outer space.

Volcanoes exploded upward—powered by the constant $E=mc^2$ derived heat beneath—and that led to something of a continuous conveyer belt from deep underground. Key trace elements were pushed up into the air, helping produce our fertile soil; great clouds of carbon dioxide were carried upward as well, creating a greenhouse effect in the young planet's atmosphere, and further ensuring the surface warmth needed for life. Where the frictional heat generated by the atoms blasting apart in accord with $E=mc^2$ was especially concentrated, deep-sea volcanoes could billow up even through thousands of feet of cold ocean water—which is how the Hawaiian islands lifted above the Pacific waves.

Fast-forward several billion years, and mobile chunks of carbon atoms emerged (in other words, us!) to wade through low-flying clouds of star-created oxygen, stir caffeine-dense liquids of Big Bang hydrogen atoms, and read about how they came to exist. For we live on a planet where $E=mc^2$ is constantly at work in the technology around us.

Atomic bombs were one of the first direct applications. At the start there were just a handful, laboriously created in the labs of the Manhattan Project, but soon there were many more, as a great infrastructure of factories and scholarships and research institutes became established after Hiroshima. Several hundred atomic or hydrogen bombs were built and ready by the end of the 1950s; today, even well after the Cold War, there are many thousands. To create them there were hundreds of open-air tests over the years, spraying immense gushes of radioactive particles into the stratosphere, there to float to every location on the planet; becoming a part of the bodies of every person alive.

Nuclear submarines were created, with radioactively exploding elements sequestered inside; pouring out heat that spun the turbines. They were fearsome weapons, yet thereby allowed a curious stability in the most dangerous phases of the Cold War. The previous generations of submarines, from World War II, had been unable to spend much time at battle stations. Cruising on the surface, World War II submarines might just manage to travel at the 12 mph of a person on a bicycle; taking the safer route, underwater, they moved at the 4 mph of a person walking. Once they'd crossed half the North Atlantic or the Pacific, they'd used up so much fuel that they quite soon had to engage in difficult wartime refueling, or turn around and trundle back. With nuclear-powered engines, it was different. Russian and American submarines could get into firing range, and then stay there for weeks or months on end— a dangerous standoff, but one at least making the other side very cautious about any moves that might provoke these hidden vessels to launch their missiles.

On land, huge electricity-generating stations were built, using the high-speed frictional heat of $E=mc^2$ to power up generating turbines. It's not the most sensible of energy choices, for even nonnuclear explosions at the generating stations can be pretty terrifying—and nothing deters corporate financial officers as much as the phrase "unlimited liability"; the radioactive walls and radioactive cement base and radioactive residual fuel from every such generator are a lot of liability to be disposed of. In France, however, the government assumes those charges, and doesn't allow court cases against the industry: about 80 percent of the country's electricity is nuclear. When the Eiffel Tower is lit at night, the electricity comes from a slower reenactment of the exploding ancient atoms that took place over Hiroshima.

$E=mc^2$ continues at work in ordinary houses. In the smoke detectors screwed tight to the kitchen ceiling, there's usually a sample of radioactive americium inside. The detector gets enough power by sucking mass out of that americium and using it as energy—in exact accord with the equation—that it can generate a smoke-sensitive charged beam, and keep on doing so for months or years on end.

The red-glowing exit signs in shopping malls and movie theaters depend directly on $E=mc^2$ as well. These signs can't rely on ordinary light sources, because they'd fail if the electricity went out in a fire. Instead, radioactive tritium is sealed inside. The signs contain enough fragile tritium nuclei that mass is constantly "lost," and usefully glowing energy sprays out instead.

In hospitals, medical diagnostics constantly harness the equation. In the powerful imaging devices known as PET scans (Positron Emission Tomography), patients breathe radioactive oxygen isotopes. The center of those atoms shatter apart, and streaks of energy coming from the destroyed mass are recorded as they emerge at extremely high speed from the body. The result is pinpointed readouts on tumors, blood flow, or drug take-ups inside the body—the workings of Prozac in the brain, for example, have been studied this way. In radiation treatment for cancer therapy, minuscule quantities of substances such as radioactive cobalt are aimed at tumors. As the unstable cobalt nuclei break apart, mass once again is seemingly torn out of existence, and the resultant energy is aimed with enough power to destroy cancerous DNA.

Yet other unstable varieties of carbon are constantly being formed outside the windows of passenger jets, created by incoming cosmic rays, some of which reach us from distant portions of the galaxy. We've been breathing in the stuff all our life. Hold a sufficiently sensitive Geiger counter over your hand, and it registers the telltale clicks. (What it's actually doing is "listening" to tiny miniatures of Einstein's 1905 equation. Every click of the Geiger counter is a mark that one or more operations of $E=mc^2$ has taken place, as the unstable nucleus of that new carbon atom plops out the extra neutron it gained high in our atmosphere.) But when we stop breathing—or when a tree dies, or a plant stops growing—no more fresh carbon is coming in. The clicks slowly die away.

This unstable carbon is the famous C-14. It's a clock, and its use has revolutionized archeology. Using carbon dating, labs could prove that the Turin Shroud was a medieval forgery, as some of the carbon in its flax had been running down since the fourteenth century, but not earlier. Carbon fragments could be collected from the Lascaux caverns, and Indian burial mounds, and Mayan pyramids, and early Cro-Magnon sites, and for the first time be used to date them accurately as well.

Soaring even higher, the satellites of the U.S. Defense Department's GPS navigation system create a constantly swirling tessellation beyond the atmosphere. The signals they beam down are constantly shifted out of sync by the time-distorting effects of relativity, as we saw in Chapter 7, and just as steadily have to be fixed, by programmers who adapted Einstein's insights to correct for the drift that would otherwise be created. And finally, perched most distant of all, is the exploding sphere of our sun, using the boomingly magnifying power of c^2 to warm our planet, as it has done for all the billions of years needed for this life-dense vista to evolve.

16

$$E=mc^2$$

Even though the sun is vast, it can't keep on burning forever. Heating the entire solar system takes immense amounts of fuel, even for a furnace that pumps material directly across the equals sign of $E=mc^2$. The sun's mass is now 2,000,000,000,000,000,000,000,000,000 tons, but it consumes about 700 billion tons of its own bulk as hydrogen fuel to keep the multimegaton blasts going each day. In a further 5 billion years, the most easily available portions of that fuel will be gone.

When that happens, and all that remains at the center is helium "ash," the reactions in our sun will start shifting upward a little bit, as fuel closer to the surface starts being pumped through $E=mc^2$. The outer layers of the sun will expand, and cool down just slightly enough to glow red. The sun will keep on expanding, and keep on glowing, until it reaches Mercury's orbit. That planet's rock surface will have already melted; fragments that are left will now be absorbed in the flames. Then, a few tens of millions of years later, our red-giant sun will reach the orbit of the planet Venus, and absorb it as well. But what will happen next?

Some say the world will end in fire,
Some say in ice.

Robert Frost published that in 1923, when he was pretending to be an apple farmer in Vermont. But he'd written the first draft when he'd been on the faculty at Amherst, and so had a good deal of time to read. Most science writers of the time had settled on the image, popular from the famous French naturalist Buffon's time through the late Victorian period, of a great cooling down of the universe. But others contrasted that with earlier apocalyptic images from Revelations, where fire and outpourings took over at the end.

What will happen to Earth is actually both. Any beings left alive on the surface of the Earth in the year A.D. 5 billion will see the sun get larger and larger until it fills about half the daytime sky. The oceans will boil away, and surface rocks will melt. Possibly life could migrate to other planets, or survive in deep tunnels, using technology unimagined now; possibly our planet will have long been barren when the emptying sun fills up the sky.

The sun will hold at that great size for about another billion years, as the helium ash left inside takes over the main burning: still seeming to pump mass out of existence; still producing fiery glowing energy in its place. Then it'll shrink, as the supporting struts of that glowing energy become too weak. In time so much fuel will have been emptied out of the sun that the burning will no longer be steady.

This is what will bring in the ice. As fuel pockets inside run low, the sun's surface will sink inward; shortly after, as other dispersed fuel sources get tapped, the energy output will roar higher again, and the surface of the sun will whip upward. Sonic booms are produced each time, but these are nothing like the brief crack of a single plane passing the sound barrier. At this stage, six billion years into our future, it's the final boom of the Titans.

Enough mass is blown away at each bounce upward, that within just a few hundred thousand years, there will be much less of our sun than before. What's left will be too weak to possess the same gravitational field it had before. If the Earth hasn't already been absorbed by the expanding sun, then—after 11 billion years of steady orbit—the sun's grip will let the planets go. The solar system breaks up, and Earth flies away.

One of the key insights into what happens next—and within which $E=mc^2$ is once again crucial—was first made by Subrahmanyan Chandrasekhar, a leader in twentieth-century astrophysics, whose career spanned almost sixty years. The discovery came when he was just nineteen, in the hot summer of 1930. The British Empire was in its dying days, but Chandra (the name he usually went by) was still within its dominion, and en route from Bombay to England, where he was taking up graduate studies at Cambridge.

There were storms in the Arabian Sea that August, keeping everyone in their cabins, but when Chandra recovered, he had weeks of quiet cruising before him, several sheaves of paper, and a family habit of always using spare time productively. It was even an occasion when the usual racism of the Empire had its advantages: Chandra was a Brahmin with dark skin, and although the children of some of the white passengers would try to play with him—and he'd oblige—the parents would quickly lead them away.

In the uninterrupted time at his deck chair, he be-

Subrahmanyan Chandrasekhar
AIP EMILIO SEGRÈ VISUAL ARCHIVES, CHANDRASEKHAR COLLECTION

came one of the first to realize something very odd about the objects in the sky above us. It was known that giant stars can explode, with their top portions rebounding away after they've collided with the heavy, collapsing core within. But what happens to that remnant core, after the explosion?

Chandra was a cultivated young man, well read in the literature of India and the West, and especially fluent in German. He'd studied Einstein's papers, and met a few of Germany's leading physicists, on their trips to India. He knew that the dense core of a star is under a lot of pressure, and now he began to think about the fact that pressure is a form of energy.

And energy is just another sort of mass.

Energy might be more diffuse than mass, perhaps, but as $E=mc^2$ shows, they're both just different versions of the same thing. Once again, the two sides of the equation—the "E" and the "m"—don't actually have to slip across and "turn into" one another. Rather, what the equation's really saying is that a chunk of what we call mass actually is energy: it's just that we're not used to recognizing it in that guise. Similarly, a glowing or compressed amount of energy really is mass: it just happens to be in a more diffuse form than we easily recognize as mass.

Chandra was about to glimpse the process leading to black holes. He merely had to trace this logic forward as it spiraled in an escapable catch-22. A compressed star core is under a lot of new pressure, and that pressure can be considered a sort of energy, and wherever there's a concentration of energy, the surrounding space and time will act just as if there's a concentration of mass. Gravity in the remnant star gets more intense, due to all this "mass." But that stronger gravity continues squashing what's left, so the pressure gets greater once more. Since the pressure can yet again be treated as simply more energy, then—as Chandra now saw by the tremendous insight of $E=mc^2$—it acts as yet more mass. The gravity ratchets up.

In a small enough star, the buildup of pressure is low enough for the stiff material near the star's center to resist it. But if the star is massive enough, the process keeps on going. It doesn't matter how tough the star's material is; indeed, if it's exceptionally resistant, that will soon just make it worse. For suppose a giant star could hold up under even greater pressure than expected: immense, unthinkable, trillions upon trillions of tons bearing down. Well, that extra pressure would "be" more energy, which would mean it acts just as if it had more mass, and so the gravity would get even stronger, compressing it ever more.

Regardless of how hard the substance is at the core, the inside of the star will be crushed until . . .

Until what?

Chandra had all the openness to fresh thoughts of youth, but even he had to pause now. Could he be predicting that the inside of the star would actually disappear? If he was right, then rips were opening up in the very substance of the universe! He took time off for prayers and meals; he even spent hours politely listening to a Christian evangelist, who explained to this devout Hindu why all religions from India were the work of the devil. "He was a missionary," Chandra remembered later, "but he was also . . . anxious to please. Why be rude to him?"

When Chandra resumed work in his deck chair, he realized that he couldn't actually say what would happen to the remaining substance of the star, as it poured into the hole created by this never-ending collapse. But it was known in accord with other work of Einstein that space and time near the star would be strongly distorted by its presence. No light would ever escape; nearby stars that were pulled into its gravitational presence would get torn apart by what seemed an "empty" location in space.

This, along with other insights, was central to the modern concept of black holes. But once Chandra reached England, his vision was resisted by almost everyone he presented it to; often with less politeness than he'd granted the missionary. Eddington himself, the man who'd been so inspiring for Cecilia Payne, was now too old for any more such fancies. It was "stellar buffoonery," he declared. It was "absurd." But by the 1960s there was the first evidence of a star (look in the direction of the constellation Cygnus the Swan, and it's a little to one side) that spins around an area that to our telescopes seems to be entirely empty space. The only thing that would be powerful enough to do this in so small a space would be a black hole. In the center of our own galaxy, there's strong evidence of another black hole, a truly monstrous one, which has accumulated to a great size over the aeons, swallowing, on average, the equivalent of one ordinary star each year. Space-time is actually being "torn" open—as the young Chandra-sekhar had been the first to see.

Chandra tried to fight Eddington's hostility in the 1930s, but when he found that even British astrophysicists who believed he was right were scared to back him in public, he ended up leaving England. He received a kinder welcome in America, and in an association with the University of Chicago went on to decades of work—culminating in his Nobel Prize in 1983, over a half century after that Arabian Sea voyage—which proved central in understanding what's in store for us next.

Six billion years from now, if Earth is flung loose from the fuel-emptied sun, any survivors or sensing devices left on our planet's surface will see a horizon darker than today's night sky. For the stars themselves will have used up their fuel and be dying out: the most fiery ones first, then the rest.

Earth's flight won't be stable through this darker expanse. Our Milky Way is already on track to collide with the Andromeda galaxy, and in several billion years, about the time of Earth's escape or immolation in the solar system, the great collision should finally happen. The spaces between stars are so great that most of the dimmed suns will just slowly pass between each other, without direct impact, but the turbulence will be enough to shift an escaped Earth's trajectory once more.

If Earth slingshots inward, then in a few tens of millions of years we will be within range to be absorbed by the giant black hole at the galaxy's center. If we get slingshot outward, however, the end will simply be delayed. By 10^{18} years from now (1 followed by eighteen zeroes, or 1,000,000,000,000,000,000 years from now), all galaxies are liable to have emptied out because of such collisions. The black holes in the centers of the galaxies will slowly travel on their own, sucking mass and energy from the universe wherever they contact other objects. If it's another black hole that they randomly impact, then they simply merge, to become an even larger devourer. A few hours after coming within range of one of these, Earth and any distant descendants on it will be taken out of existence.

By 10^{32} years into the future, protons themselves might have begun to decay, and gradually very little of ordinary matter will be left. The universe will be composed of a greatly reduced category of things. There will be electrons of the sort we're used to, with a negative electric charge, and there will be curious antimatter versions of electrons, with a positive charge, and along with neutrinos and gravitons there will be the swollen black holes, and even a cooled remnant of photons left over from the first seconds of creation, still traveling at their eternal 670 million mph speed after all these ages.

It doesn't end there, for given enough time even black holes can evaporate. Everything they engulfed will be released back—not in any recognizable form, but as an equivalent amount of radiation.

The universe will have ended up in a state curiously transposed from what it was at the start. For in the very first moments of creation, long before the sun was formed, the universe was immensely dense, immensely "concentrated." That great density meant that large amounts of radiation were "pushed" along $E=mc^2$, from the "E" side to the "m" side. The ordinary matter we're familiar with took shape out of pure energy, ultimately creating the stars and planets and lifeforms we know. But now, near the end of time, over 10,000,000,000,000,000,000,000,000,000,000,000,000, 000,000,000,000,000,000,000,000,000,000,000,000, 000,000,000,000,000,000,000,000 years into the future, it's different. Everything is far more dispersed, far more diffuse.

What will exist then will be spread over distances we cannot imagine. The rush of activity of early epochs will be over. That was just an interlude in the final history of the universe. Now, mass and energy only very rarely transform into each other. There is a great stillness.

The work of Einstein's equation is done.

$E=mc^2$

What Else Einstein Did

It wasn't actually $E=mc^2$ and his other work from 1905 that first made Einstein famous. If that were all he had done, his name would have become recognized within the specialized community of theoretical physicists, but probably not otherwise known to the public. In the 1930s he would have been just another distinguished refugee: living a quiet life perhaps, but in no special position to sign a letter warning of atomic dangers, which could be delivered to FDR in 1939.

It didn't turn out like that of course. Something else happened that built on $E=mc^2$ but went further—and ended up making him the most famous scientist in the world.

What Einstein published in 1905 only covered cases where objects are racing along smoothly, and gravity, with its accelerating pull, doesn't play much of a role. $E=mc^2$ is "true" in those cases, but will it hold true even if you get rid of those restrictions? That limitation and others had always troubled Einstein, and in 1907 he got the first hint of a wider solution: "I was sitting on a chair in my patent office in Bern when all of a sudden a thought occurred to me. . . . I was startled."

He later called this "the happiest thought of my life," for a few years later, in 1910, it led to his reflecting on the very fabric of space, and how it was affected by the mass or energy of objects at any one location in it. The work took several years, partly because although Einstein was in a league of his own in physics, he was only fair in mathematics. It wasn't quite as bad as he once described to a junior high school student in America, when he wrote her, "Do not worry about your difficulties in mathematics. I can assure you that mine are still greater." But it was enough to justify Hermann Minkowski's lament, when he saw the early drafts of Einstein's efforts: "Einstein's presentation of his subtle theory is mathematically cumbersome—I am allowed to say so because he learned his mathematics from me in Zurich."

To help him with the math, though, Einstein had his old friend from university days, Marcel Grossman, the one who'd loaned him crib sheets when they were undergraduates. (Grossman was also the friend whose father had written the letter getting Einstein the patent office job.) Grossman sat with Einstein for long hours, to explain what tools from recent mathematics he might use.

What Einstein's "happiest thought" of 1907 led him to was the idea that the more mass or energy there was at any one spot, the more that space and time would be curved tight around it. It was a far more powerful theory than what he'd come up with before, for it encompasses so much more. The 1905 work had been labeled "special" relativity. This now was general relativity.

A small, rocky object, such as our planet, has only a little bit of mass and energy, and so only curves the fabric of space and time around it a bit. The more powerful sun would tug the underlying fabric around it far more taut.

Sun   Earth

Examples of warped space-time.
MODIFIED FROM A DIAGRAM IN *THE SCIENCES* BY
JAMES TREFIL AND ROBERT M. HAZEN
(NEW YORK: JOHN WILEY & SONS, 1998)

The equation that summarizes this has a great simplicity, curiously reminiscent of the simplicity of $E=mc^2$. In $E=mc^2$, there's an energy realm on one side, a mass realm on the other, and the bridge of the "=" sign linking them. $E=mc^2$ is, at heart, the assertion that Energy = mass. In Einstein's new, wider theory, the points that are covered deal with the way that all of "energy-mass" in an area is associated with all of "space-time" nearby, or, symbolically, the way that Energy-mass = space-time. The "E" and the "m" of $E=mc^2$ are now just items to go on one side of this deeper equation.

The entire mass-loaded Earth rolls forward, automatically following the shortest path amidst the space-time "curves" that spread rippling around us. Gravity is no longer something that happens stretching across an inert space: rather, gravity is simply what we notice when we happen to be traveling within a particular configuration of space and time.

The problem, though, is that it seems preposterous! How can seemingly empty space and time be warped? Clearly that would have to occur, if this extended theory, which now embedded $E=mc^2$ in its wider context, were to be true. Einstein realized that there could be something of a test—some demonstration that would be so clear, so powerful, that no one could doubt that this wild result he'd come up with was right.

But what could that be? The proving test came from the heart of the theory, that diagram of a warp in the very fabric around us. If empty space really could be tugged and curved, then we'd be able to see distant starlight "mysteriously" swiveled around our sun. It would be like watching a bank shot in billiards suddenly take place, where a ball spins around a pocket and comes out with a changed direction. Only now it would occur in the sky overhead, where nobody had ever suspected a curved corner pocket to reside.

Normally we couldn't notice this light being bent by the sun, because it would apply only to starlight that skimmed very close to the outer edge of our sun. Under ordinary circumstances the sun's glare would block out those adjacent daytime stars.

But during an eclipse?

Every hero needs an assistant. Moses had Aaron. Jesus had his disciples.

Einstein, alas, got Freundlich.

Erwin Freundlich was a junior assistant at the Royal Prussian Observatory in Berlin. I wouldn't say he had the worst luck of any individual I've read about. Possibly there was someone who survived the *Titanic*, and then decided to try a ride on the *Hindenburg*. But it's probably pretty close. Freundlich was going to make his career, he decided, by shepherding the great general relativity equations forward, and performing the observations that would prove Professor Einstein's predictions were right. He was very generous about this—in the way that Lavoisier had been generous in letting his wife help him watch metal heat and rust. As a special honeymoon treat, Freundlich brought his new bride to Zurich in 1913 just so she could be there as he discussed stellar observations with the renowned professor.

An eclipse was predicted for the very next year, in the Crimea, and Freundlich prepared everything in detail. He even carefully arrived in the Crimea two months early, in July 1914. It was probably the worst possible place for a German national to be. War was declared one month later. Freundlich was arrested, put in prison in Odessa, and had all his equipment taken away. He finally got out in a prisoner exchange for a group of Russian officers who'd been arrested in Germany, but by then the eclipse had come and gone.

He didn't give up. In 1915, back in Berlin, Freundlich decided he could help Professor Einstein by measuring the way light got bent near distant binary stars. In February he had results that backed up the new theory, and Einstein began to spread the good news in letters to his friends. Four months later, though, Freundlich's colleagues at the observatory found he'd estimated the mass of the stars all wrong, and Einstein had to take it all back. For most people (as Freundlich's young wife perhaps tried to explain) that would have been enough, but Freundlich resolved to try yet again. Why didn't they try measuring how much distant starlight got deflected near the massive planet Jupiter; the one that the great Roemer himself had so persuasively used to resolve a scientific problem in an earlier era? Freundlich proposed it to Einstein. Einstein liked his earnest young helper, and in December he wrote to Freundlich's director at the Prussian Observatory, suggesting that he be allowed to try this.

It would have been less painful just to have sent him back to the Crimean prison. Freundlich's superior was furious that anyone would dare to interfere. He threatened to fire Freundlich, insulted him in front of his colleagues, and made sure that he never, ever was allowed to get his hands on the equipment that could be used to test the prediction near the orbit of Jupiter.

But that didn't matter. Freundlich was hopeful again. A great new eclipse expedition was being planned, for 1919. If conditions allowed international travel, he'd finally be able to prove what he could do.

In November 1918 World War I ended. There were no obstacles to a German national traveling now! It's not recorded what Freundlich felt as the great expedition set out, but we know exactly where he was when the results came through. He read it in the newspaper, back in Berlin.

He hadn't been invited along.

In fact, it was a cool Englishman we've already met who led the team. Arthur Eddington wore small metal-rimmed glasses, was medium height and barely medium weight, and spoke in sentences that tapered off whenever he had to pause for thought, which was fairly often. This of course meant in the good English manner that under his meek exterior there beat a soul of wild determination. By the time Chandra encountered him in the 1930s his personality had hardened, but at this time, in the period of World War I, he had the energy of a young man.

On May 29 of each year the sun is positioned in front of an exceptionally dense group of bright stars—the Hyades cluster. That wouldn't usually help anyone, for without a solar eclipse occurring on that particular date, there would be no chance to see how that rich field of stars gets their light bent around the sun. The glare from the daytime sun would overwhelm that small effect. But in 1919 there was going to be an eclipse, precisely on May 29. As Eddington innocently noted: "Attention was called to this remarkable opportunity by the Astronomer Royal [Frank Dyson] in March 1917; and preparations were begun. . . ."

What Eddington neglected to mention was that he would have been thrown into prison if he didn't go. For as a Quaker, Eddington was a pacifist, and as a pacifist, in the middle of World War I England, one of the rough prison camps in the Midlands was in store. The soldiers guarding the pacifist camps were often recently back from the front—or embarrassed that they themselves hadn't seen service there, which could be worse. Conditions were rough. There was steady abuse and beatings; a number of deaths.

Eddington's colleagues at Cambridge didn't want him to go through this, and tried to arrange for the War Department to defer him, as being important for the nation's scientific future. A letter confirming this was sent to him, from the Home Office, which he only had to sign and send back.

Eddington knew what was in store in the prison camps, but being a pacifist isn't the same as being a coward, as the actions of many Quakers years later in the American civil rights movement showed. Eddington signed the letter, since that was only fair to his friends, but then he also added a postscript, explaining to the Home Office that if he wasn't deferred on grounds of scientific usefulness, he'd still ask to be deferred as a conscientious objector. The Home Office was not impressed, and began proceedings to send him to one of the prisons.

Arthur Eddington
AIP EMILIO SEGRÈ VISUAL ARCHIVES

This is the point at which the Astronomer Royal, Frank Dyson, called attention to the remarkable eclipse opportunity. If Dyson could get Eddington to arrange the expedition, could Eddington still be deferred, despite that postscript? Dyson's work was relevant to navigation, and so he was close to the admiralty. The admiralty had a word with the Home Office. Eddington was free . . . so long as he led that expedition. They had two years to prepare.

It rained during the expedition, of course, but this is only what you'd expect on an island off the African coast, just north of the Congo, where Eddington ended up. But remember, Freundlich wasn't with Eddington. The rain cleared, and Eddington got two good plates. Most of the developing would have to be done in Britain, however, and no one would know the result for several months.

Afterward, Einstein tried to pretend that he hadn't been bothered by the delay. But by mid-September, still having no word, he wrote to his friend Ehrenfest, asking, with overelaborate casualness, if perhaps *he'd* heard anything about the expedition? Ehrenfest had good connections with the British. But no, he knew nothing. He wasn't even sure if Eddington had made it back.

In fact, Eddington had been back at Cambridge for several weeks now, but his photographic plates were a mess. They'd been carried by ship to West Africa, then kept in tents on a humid island, then carried into the rainstorm at the start of the eclipse, handled in and out of the camera, then brought back to the tents, and finally shipped by ocean steamer once more. The physical separations Eddington was looking for, in the movement of the distant stars, were going to be measured in tenths of a second of arc. On the small photographic plates, that came to barely a millimeter. (A thick pencil line is about one millimeter. If you have very good vision, you can just make out dust motes 1/20 of a millimeter across.) Eddington had micrometers to help, but Einstein would only be right if these tiny displacements were exactly as predicted, and so far, Eddington couldn't see them clearly enough to be sure if they were. The emulsion from the West Africa plates had become so jellylike in all the heat and transportation that if Eddington was honest he might never make out the necessary detail.

No one at Cambridge wanted to give up, though, for Einstein's was such a sweet theory. It was tremendous to think that the great tumbling ball of the sun was crashing down on the very fabric of space and time, sagging it so much that distant starlight started veering sideways as it got caught in the bend. Nor was it just the "traditional" mass of the sun that would be doing this. The 1905 equation entered in also. All the heat and radiation blasting out of the sun—all that "energy"—was acting as an additional form of "mass." It added to the bulk of the sun as well. (This was at the heart of what Chandra would build on, in his later sea voyage of 1930.)

Luckily, the British Empire had its traditions, and one of the prime ones was that something always went wrong. Explorers, conquerors, younger sons and even metal-eyeglassed Quaker astronomers had learned that lesson: picking it up from a lifetime of hearing about one imperial expedition after another.

And that's why Eddington had sent out a second team—an entire duplicate crew—to be sure he proved Einstein's prediction.

This second crew had a different telescope, and had been sent to a different continent (they'd been in northern Brazil), and they even had a different mechanical drive for the telescope. It was all in the finest tradition of spreading the odds, and it worked. Once the Brazil team's plates came back, and a special oversized micrometer had been built to fit around their larger plates, and Eddington and the others had measured and remeasured, the congratulatory telegrams started bursting out. Bertrand Russell, who had recently been a Fellow at Trinity, now received a message from his old friend Littlewood: "Dear Russell: Einstein's theory is completely confirmed. The predicted displacement was 1".72 and the observed .1"75 +/- .06."

The celebration was in style. The Royal Astronomical Society was invited to a joint session with the Royal Society on November 6, 1919, in the great room at Burlington House, on Piccadilly. Scientists came in from Cambridge and elsewhere to the stations at King's Cross and Liverpool Street; cabs were taken; nonscientists who'd heard something momentous was to be announced arrived as well. A visitor described the evening: "There was dramatic quality in the very staging:—the traditional ceremonial, and in the background the picture of Newton to remind us that the greatest of scientific generalizations was now, after more than two centuries, to receive its first modifications."

Dyson spoke, and Eddington spoke—there's no record if any narrow-eyed parole officer from the Home Office was in the room—and then the elderly chairman stood up to speak:

> This is the most important result obtained in connection with the theory of gravitation since Newton's day, and it is fitting that it should be announced at a meeting of the Society so closely connected with him. . . .
>
> If it is sustained that Einstein's reasoning holds good . . . then it is the result of one of the highest achievements of human thought.

With World War I just over, these findings were wondrous. God may have seemed lost after the trenches, but now order had been divined in the cosmos. Even better, a German and an Englishman working in harmony had found it. Royalty and generals and political leaders and even artistic figures who'd made their reputation under the old regime—the regime that had led to the slaughters of World War I—were discredited. The category of "people to respect" was nearly empty. Einstein, instantly, was the greatest media celebrity on the planet. Headlines in the *New York Times* for November 10, 1919, announced:

"Light All Askew in the Heavens: Men of Science More or Less Agog Over Results of Eclipse Observations."

and

"Einstein Theory Triumphs: Stars Not Where They Seemed or Were Calculated to Be, but Nobody Need Worry."

This meeting was also when the rumors began that only a dozen people could understand what it all meant. The *New York Times* did have a few knowledgeable science writers, but they were in New York. The London bureau was handed the story, and Henry Crouch was asked to cover Burlington House. In the history of inappropriate assignments this is at the Lyman Briggs level. Crouch was a good journalist in the sense that he knew you had to make a story interesting. He was somewhat less good, however, in having the slightest clue what was going on here—Crouch was the paper's golfing specialist.

But he was also a *Times* man through and through, and nothing like a simple lack of knowledge was going to hold him back. He kept on filing, and the headline writers pulled out the key parts of his story:

"A Book for 12 Wise Men: No More in All the World Could Comprehend It, Said Einstein When His Daring Publishers Accepted It."

He made that up. Einstein wasn't writing a book, there were no publishers involved—daring or not—and most of the physicists and astronomers attending understood easily enough what the meeting was about. Crouch had started the theory off on its track record of poor public comprehension, from which it never entirely recovered.

That only added to its fame. In almost all religions, there's a powerful difference between a priest and a prophet. A priest merely stands below an open hole in the sky, and lets the truth that's normally kept hidden up there come pouring down. (Press secretaries and nuclear technicians are examples.) A prophet, however, is someone who manages to journey up through that opening. They are individuals who can venture to that Other Side, before returning back to ordinary life, here with us on Earth. As a result, we'll try to glimpse, in the expression on their face, or in the potent equations they've plucked and brought back down, what things are like up there, in that higher realm, which so many of us believe in, but know we'll never get to visit directly.

Martin Luther King Jr. and Nelson Mandela have been considered such prophets, carrying down a vision of racial harmony, their words spreading afterward with a power that came from the feeling that they had originated from that higher source. In post–World War I Europe, Einstein's findings were received with the veneration King's or Mandela's words would be granted later. And since very few people understood Einstein's work at first, all the feeling it suggested—all the desire for transcendence and for knowledge from Einstein's divine library—would soon be shifted onto images of Einstein himself. Perhaps that's why people were attracted to photos of him that had a distinctive, sadly bemused look. They matched the later, most powerful photos of Martin Luther King, where he too seemed to be sadly seeing something greater than ordinary mortals could.

Einstein tried to push back some of the fame. He

called the exaggerating newspaper accounts an amusing feat of imagination. Two weeks after the public announcement, he wrote in the London *Times* that although the Germans were proudly calling him a German, and the English were proclaiming him a Swiss Jew, if his prediction ever came to be shown false, the Germans would call him a Swiss Jew, and the English would call him a German. In fact, he got it wrong: his astronomical prediction and the 1905 equation both stayed true, but English anti-Semites such as Keynes still scorned him ("a naughty Jew-boy, covered with ink"), and with the rise of Hitler the German government not only called him a Jew but supported the calls to have him killed. After leaving the Continent, and trying England, he ended up in America for the rest of his life: in 1939 signing the letter to President Roosevelt, which, albeit indirectly, helped lead to the atomic bomb; otherwise just living a quiet professorial life at 112 Mercer Street, in Princeton, New Jersey.

He never especially liked the Ivy League snobbery of Princeton ("this village of puny demi-gods upon stilts," as he described it to a European friend). There were giggling bobby-soxers; the occasional gaping tourist; at the Institute for Advanced Study—a two-mile walk from his home, which he took regularly—younger scientists kept a surface politeness, but he knew that many disparaged him behind his back as someone too old to be useful.

That alone didn't seem to bother Einstein. His goal, as always, was simply to see what had been intended for our universe by The Old One. What he had scribbled in his now-yellowing manuscripts decades before, as well as the new equations he was constantly working on now—trying to create a theory that would unify in a clear and predictable way all the known forces in the universe—still seemed, to him, the best possible track forward.

What did hurt him were different reminders of how things had worked out. One, almost too horrible to think of vividly, was implicitly brought up each time he encountered Oppenheimer, his institute's head, who'd led the Manhattan Project that had demonstrated that $E=mc^2$—despite Einstein's lack of involvement—could be turned into vast fields of death in Hiroshima and Nagasaki. "Had I known that the Germans would not succeed in producing an atomic bomb," Einstein once told his longtime secretary, "I would never have lifted a finger. Not a single finger!"

Then, as the years went on, there was the increasing feeling of his own powers fading away. An unintentionally tactless young assistant once asked him about this. Einstein explained that it was more difficult now to judge which of his ideas were worth pursuing—a great contrast with his younger years, when he'd been superb at identifying the key issues in a field. "Discovery in the grand manner is for young people," he'd once told a friend, ". . . and hence for me a thing of the past."

He settled into an old man's daily routine, in his simple clapboard suburban house on Mercer Street. His sister, Maja, was with him in America by now. She suffered a severe stroke in 1946, and virtually every evening after that, for the six years till her death, Einstein would drop whatever work he was doing and go to her room, where he would read aloud to her for hours. Before that, most days, there were the mock-chiding rituals with his housekeeper; the saddened disregarding of reminders of his mentally disturbed second son; sometimes visits from a friend with whom he'd enjoyed playing Bach's Double Concerto, as well as the violin parts in Purcell's or Handel's Baroque trios. But there

Albert Einstein
LIBRARY OF CONGRESS. EMILIO SEGRÈ VISUAL ARCHIVES

were also the moments, settled comfortably in his up-stairs study, when his steadily penciled pages of symbols lifted him back into his past, to the time when anything had seemed possible.

And the works—of the divine library that he was convinced awaited—could once again be read.

Follow-up of Other Key Participants

When MICHAEL FARADAY took over Davy's position at the Royal Institution, he moved permanently into the Royal Institution with his wife. He continued to make major discoveries well into his fifties, but despite many requests, he never took on a personal student.

After the execution of ANTOINE-LAURENT LAVOISIER, his remains were carted out of Paris, passing through one of his new tollgates that had survived the 1789 attacks. A few months after Lavoisier's death, the body of the man who'd ordered the executions, Marat's colleague Robespierre, was carted through the same gate, and placed in the same common grave. It was a converted wasteland called "Errancis" ("maimed person"), which they now shared. Several fragments of the solid tollgates from the GENERAL FARM wall, which Lavoisier had ordered built, can be seen to this day, in the Parc Monceau, and near the metro exit at Denfert-Rochereau.

A few months before Lavoisier's arrest, a young woman, Charlotte Corday, called at the apartment of JEAN-PAUL MARAT, asking to see him. His guards refused, but when she insisted she had news about dangerous political opponents, he overruled them, and had her let upstairs. Since Marat had skin complaints that forced him to spend much of the day in a bathtub, it was from that position that he greeted her, discovered that the political opponents were members of her family (whom he had ordered killed), and then saw her step forward, knife out. She stabbed him to death, in an assassination later immortalized by the society painter David.

Since MARIE ANNE PAULZE had been only thirteen when she married Lavoisier, she was just thirty-five when her husband was killed. Although harassed by the Revolutionary government, and her wealthy apartment emptied, she outlived most of her persecutors, and enjoyed a peaceful old age.

When he went back to Denmark, OLE ROEMER married the daughter of his ethics professor—the man who'd first brought him to the attention of Cassini's scout. Roemer ended up becoming a senior highway inspector, then mayor of Copenhagen, then prefect of police, and for several years was also the equivalent of a Supreme Court judge. In his spare time he worked on an improved device for the measurement of temperatures, which a visiting businessman named Daniel Fahrenheit thought had some merit. Roemer died in 1710, seventeen years before the British experiments that finally proved he'd been right about the speed of light.

JEAN-DOMINIQUE CASSINI outlived Roemer, and continued to promote only those astronomers who—erroneously—agreed with him that light traveled at an unmeasurable speed. The dynasty he established ran for almost two centuries, up till the fourth generation, ending with the Cassini who was forced to close down his great-grandfather's proud Observatory—the building Lavoisier had seen from his prison window.

In 1997, a spacecraft with a European Space Agency (ESA) probe on board was launched on a seven year journey to Saturn. Along its way it flew past the planet Jupiter, which Roemer had used for his epochal prediction. The spacecraft was named *Cassini*. France is a major funder of the ESA.

VOLTAIRE lived to extreme old age, writing and mocking all the way. His collected works run to over 10,000 printed pages, and did much to promote the Revolution, which began just a few years after his death. He never published significant commentaries on science after du Châtelet's death.

The manuscript that **EMILIE DU CHÂTELET** finished in her final days—*Principes Mathématiques de la Philosophie Naturelle*—became a great success in the scientific circles of its time. A first edition can be viewed at the Bibliothèque Nationale in Paris. The Château de Cirey ended up shuttered and abandoned during the Revolution, but was later refurbished. Her first son never lived to see that, having become ambassador to Britain under Louis XVI, which led, after his return to France, to his arrest and subsequent death by the guillotine. "If I were King," du Châtelet once wrote, ". . . women would be worth more, and men would gain something new to emulate."

HENRI POINCARÉ lived for seven years after Einstein's 1905 publications, still unreconciled to the fact that outside of France he wasn't recognized as a founder of relativity. In his final years he wrote eloquent, thoughtful essays on creativity. He also ensured that no one who wanted to work on Einstein's theories could be promoted in France.

MILEVA MARIĆ-EINSTEIN continued looking up to her husband, even as he started an affair and their marriage broke apart. When he promised to give her any future Nobel Prize money as a divorce settlement, she saw nothing unusual in assuming that he would win it. (In 1922, when he did get the prize—though not for his theory of relativity, as the Swedish Academy was still not entirely convinced it would prove itself—he promptly transferred to her the substantial prize money as promised.)

She never remarried after the divorce, and having missed her chance to retake her final university exams (her grades had been just slightly too low to get a teaching job), she never found a significant career. Although her first son ended up an engineering professor at Berkeley, she became exhausted caring for their second son, who was in and out of mental institutions his whole life. She died in Zurich in 1948, increasingly depressed and alone.

MICHELE BESSO, Einstein's closest friend from his Bern years, with whom the ideas of special relativity were first talked over, had a rich home life, and a successful career as a mechanical engineer. Even in the 1950s, when he and the now twice-married Einstein were old, their correspondence continued and even became more frequent. After Besso's death, in early 1955, Einstein wrote Besso's family: "The gift of leading a harmonious life is rarely joined to such a keen intelligence, especially to the degree one found in him. . . . what I admired most about Michele was the fact that he was able to live so many years with one woman, not only in peace but also in constant unity, something I have lamentably failed at twice. . . ."

Despite the bowling balls and child's hoe that had been thrown her way as a child, MAJA EINSTEIN became her big brother's closest friend. In 1906 she moved to Bern in part to be close to him, and ended up taking a doctorate (in Romance languages) from the university there, an extremely rare achievement for a woman at the time. When Einstein began teaching at that university she—and Besso—regularly attended some of his first classes, so that the authorities would be less likely to notice how few other students Einstein was then getting.

ERNEST RUTHERFORD died suddenly in 1937, following an intestinal rupture, which was possibly linked to overvigorous gardening he was doing at his weekend cottage. His final words were for his wife to be sure to arrange that scholarship funds be sent to Nelson College, in New Zealand, which is where he had received the schooling that raised him from rural poverty, and prepared him for his own scholarship to England. THE CAVENDISH LABORATORY he left behind never again achieved the same preeminence in nuclear research. In time a new director shifted it increasingly toward biology. This included welcoming a young American, James Watson, who it was thought might work well with the physics-trained Francis Crick, in using the Cavendish's resources to try investigating the structure of DNA.

HANS GEIGER, Rutherford's young man who'd had the knack of making such useful radiation counters, returned to Germany, and soon assumed senior academic positions. His years in England had, however, little effect in making him a believer in tolerance or freedom. He was one of the most active of senior German physicists in supporting the rise of Hitler, and welcoming students with swastikas. He turned against his Jewish colleagues, including ones who'd helped him over the years; as Hans Bethe and others have noted, he seemed to enjoy coldly turning down any of their requests for aid in obtaining foreign posts.

SIR JAMES CHADWICK was holidaying with his family on the Continent when the German invasion of Poland began in 1939, and although he was assured by his hosts that there was no chance of being caught behind enemy lines, he brought his family back to England with remarkable alacrity. Having stood up to Oppenheimer enough to impress General Groves, he was brought into the centers of power in Washington, and turned out to be one of the most effective administrators of the Manhattan Project. He lived into the 1970s, but had been so distressed by what the bomb explosions could lead to that "I had then to start taking sleeping pills. It was the only remedy. I've never stopped since then. It's 28 years, and I don't think I've missed a single night in all those 28 years."

ENRICO FERMI had been at ease with virtually everyone he worked with in Italy, and repeated the process in America. He worked hard to master American colloquialisms, and admitted failure in his Americanization efforts only when it came to clearing the lawn of his first suburban house of crabgrass—for did it not, he and his wife inquired, have as much of a right to grow there as anything else?

His participation in the Manhattan Project was central to its success, but as with a number of the participating scientists, he was hit by cancer when he was still only in middle age. He was notably calm in the hospital room during his last few months. When the Hindu Chandrasekhar came in, unsure what to say, Fermi put him at ease by asking, with a smile, if Chandra could tell him if he was going to come back as an elephant the next time.

America's largest high-energy physics research center is located about 30 miles southwest of Chicago. It is called FERMILAB.

OTTO HAHN received the Nobel Prize for the work that Lise Meitner had led him toward. Instead of explaining that this was a mistake, and that she should have been honored, even if only jointly, he began to write her out of the story. In his first postwar interviews he started saying that she had merely been a junior research assistant; later he pretended (believed?) that he'd barely heard of her at all.

For many years, the workbench Meitner had used in Berlin, with all the devices she'd accumulated for the key experiment, was on display in the Deutsches Museum in Munich. It was labeled the *Arbeitstisch von Otto Hahn*: The workbench of Otto Hahn.

As a mark of Hahn's fame, when a new chemical element, number 105 was created, it was named HAHNIUM. In 1997, however, the name vanished from the the Periodic Table; the new element was officially relabeled Dubnium in honor of the Russian city where it had first been created.

FRITZ STRASSMANN was disappointed at Hahn's antics, and refused the 10 percent of the Nobel Prize money that Hahn later offered him. He kept his liberal sympathies even in the midst of the war, hiding the Jewish pianist Andrea Wolffenstein for several months in his Berlin apartment—for which he was later honored at the Holocaust memorial Yad Vashem, in Jerusalem. After the war Strassmann wrote to Meitner, asking her to return to Germany, but noting that he'd understand if she didn't.

LISE MEITNER was hurt at what her lifelong partner Hahn did to her, but put it down to his desire to suppress everything about the recent German past. She left Stockholm for Cambridge, England, and in the 1960s could be seen as a slender very old woman, browsing in the bookshops. Into her mid-eighties she kept a notebook of questions to ask her young nephew. These included topics in current theoretical physics, as well as perplexing vocabulary words such as *highfalutin'* and *juke box*. She died in relative obscurity in October 1968, a few weeks after the world-famous Hahn.

In the 1970s feminist scholars began to reexplore her career. When a new chemical element, number 109 in the Periodic Table, was created in 1982, it was named MEITNERIUM.

The young nephew, ROBERT FRISCH, managed to get out of Denmark before the German army invaded. Successfully reaching England, he was barred from classified work on radar because he was an enemy alien, and so had time for the computations that showed that much less uranium than suspected would be enough for a bomb. This was the basis for the classified report that jump-started the U.S. bomb project when it was finally brought out of Lyman Briggs's safe.

Frisch played an important role at Los Alamos, though by March 1945 he was back in Cambridge, where he happened to be at the Cavendish laboratory when the young Fred Hoyle came by, in need of some listings of nuclear masses for an idea he'd had about the way elements were formed inside stars. Frisch supplied them.

After the war, with his first name now "Otto," Frisch continued to be a firm anglophile, though he always retained a suspicion that "the weather" was something which had only recently arrived in Britain, this being the only reasonable explanation as to why the populace commented on it so frequently. To his great pleasure, in 1947 he was offered a named professorship at Cambridge—so allowing him to share in the tradition as had an earlier immigrant, Ernest Rutherford.

As soon as the bombs to be used against Japan were delivered, J. ROBERT OPPENHEIMER went back to being as sarcastic as ever, suddenly addressing the staff who remained at Los Alamos as second raters. He also applied his sharp tongue to Lewis Strauss, head of the new Atomic Energy Commission (AEC), as well as to Edward Teller, which meant that he had serious enemies when a witch-hunting AEC committee investigated his 1930s attendance at left-wing parties, as well as his moral reticence about the hydrogen bomb. In 1954 he was purged from all government service.

LESLIE GROVES always kept a soft spot for Oppenheimer. Retired from the army, and an executive at Remington Rand, he refused to condemn Oppenheimer wholeheartedly (as most other army staff did) at the 1954 hearings. Groves always held that Oppenheimer was "A real genius. . . . Lawrence is very bright, [but] he's not a genius, just a good hard worker. Why, Oppenheimer knows about everything. He can talk to you about anything you bring up. Well, not exactly. I guess there are a few things he doesn't know about. He doesn't know about sports."

Using material from Lawrence's lab, EMILIO SEGRÈ had become the first person to create the element technetium. He also managed to stay at the Berkeley Lab long enough to become the codiscoverer of plutonium, the element used in the Nagasaki explosion. At the reduced salary Lawrence gave him, there had been no chance of bribing any consular officials to get his elderly parents out of Italy. His mother was captured during a Nazi manhunt in October 1943, and murdered soon after that; his father, who had been safely hidden in a papal palace, died of natural causes the next year.

When the war was over Segrè went to his father's tomb, scattering a small sample of technetium from Lawrence's lab over it: "The radioactivity was minuscule, but its half-life of hundreds of thousands of years will last longer than any other monument I could offer."

As soon as Denmark was liberated, GEORGE DE HEVESY went back to the jar of strong acid in which he'd dissolved the Nobel gold medals at Niels Bohr's Copenhagen institute, and simply precipitated them back out. The Nobel foundation then recast them, and they were returned to their rightful owners. When de Hevesy had first dissolved them he'd only just recovered from a full-fledged midlife crisis, convinced that at age fifty he was past the age for fresh invention. The recovery was quite complete, for soon he had a Nobel medal of his own, awarded for work he did—at an age when most physicists' creativity is long gone—on radioactive tracers.

All laureates are offered Swedish citizenship, but de Hevesy was one of the few who took that up, settling in Stockholm for the rest of his long life. In the 1960s, he could sometimes be seen strolling in La Jolla, California, an erect elderly man, happy visiting with his American grandchildren, telling them what he remembered of life growing up in the 1880s in a baronial palace in Hungary.

ERNEST LAWRENCE came out of the war in triumph, and succeeded in raising more and more funds, and building larger and larger machines, until finally he proposed a cyclotron that violated the special theory of relativity, and so was physically impossible. None of his young men would dare to explain that to him, however, and the failure of his efforts to get it to work ended up wrecking his health. A little before he died, in 1958, he told a group of graduate students at the University of Illinois: "Why, fellows, you don't want a big machine. There's too much emphasis these days on sheer size for its own sake."

WERNER HEISENBERG became the grand old man of German science, and after a brief six-month internment in the luxury of a grand country house in Cambridgeshire, England, was soon respected worldwide as a sage and philosopher. He rarely spoke of the war, but when he did, would give the impression through hints and nodding gestures that he had been able to make a bomb all along, but had willfully led the research in the wrong direction, to keep the Nazi government from getting the weapon.

Heisenberg never realized he was being recorded at the Cambridgeshire country house.

HEISENBERG: *Microphones installed? [laughing] Oh no, they're not as cute as all that. I don't think they know the real Gestapo methods; they're a bit old-fashioned in that respect.*

But when the recordings were released a half century later they proved Heisenberg's cover story false. There was a fine justice in Heisenberg and the others being sequestered there, for it was only a short distance from the other elegant country house that the British secret service kept, where the six Norwegians who destroyed his project had prepared for their mission.

Heisenberg almost hadn't survived to be captured, for the predecessor of the CIA had sent an assassin, the ex-athlete Moe Berg, against him during that final Swiss trip. Berg was planted in the audience of the seminar Heisenberg gave in Zurich. If Heisenberg showed evidence that his bomb project was on the right tracks, he would be killed. Berg had a gun, and understood some undergraduate-level physics, but the talk was too technical for him to follow. His scrawled notes from that meeting still survive in official archives: "As I listen, I am uncertain—see: Heisenberg's uncertainty principle—what to do to H. . . ." He left Heisenberg alone.

KNUT HAUKELID survived the war, despite the vast manhunt that began after his sinking of the Lake Tinnsjö ferry. Transcripts from Heisenberg's internment finally clarified the significance of that sinking, where the equivalent of about 600 liters of concentrated heavy water had gone down. (In the following extract, Heisenberg is speaking in English, thus the imperfect grammar):

HEISENBERG: *We have tried to make a machine which can be made out of ordinary uranium. . . .*

(QUESTIONER): *With a little bit of enrichment?*

HEISENBERG: *Yes. That worked out very nicely and so we were interested in it.*

(*Pause*)

After our last experiments, if we had 500 liters more heavy water, I don't doubt that we had got the machine going. . . .

Haukelid became an officer in the Norwegian army; another member of the original commando team put Thor Heyerdahl at ease by sailing with him on *Kon-Tiki*.

The heavy water facility at VEMORK continued in operation till the early 1970s, when, having outlived its economic usefulness, it was blown up by Norsk Hydro engineers. Some of the rubble was removed by truck and train, but much was left in place, and simply paved over. Several thousand visitors walk over it each year, for the old generating station behind it has been converted into an excellent museum, and the location of the commando raid is directly under the route to the entrance.

The I. G. FARBEN company, which had taken over the plant's operation during the war, was briefly broken up by Allied authorities, after the Nuremberg trials showed its executives profiting from the purchase and subsequent death of human slaves. One of its main constituents, the BAYER company, though popularly known just for its aspirin, continued to be a major force in general chemicals worldwide.

The BERLIN AUER factories, where female slaves from Sachsenhausen had been worked to death to supply the German project with uranium oxides, almost survived intact till the end of the war. In the last few months, though, they were obliterated by Allied bombers acting on Groves's instructions, in large part to keep them from falling into Russian hands. Almost all the Berlin Auer executives avoided jail sentences, and indeed even before the war ended had been thinking of their future. American investigators found that all Europe's supplies of radioactive thorium had been purchased by an unknown buyer—it was the Berlin Auer company, which planned to use it to make white-glowing toothpaste once more.

War crime trials in Oslo after the war led to the conviction of several guards—both Germans, and Norwegian collaborators—responsible for the deaths of the surrendered BRITISH AIRBORNE TROOPS. Many of the troops had been thrown into shallow graves, with their hands tied behind their backs with barbed wire. They were disinterred for re-burial; the head of the Norwegian collaborationist govern-ment, Vidkun Quisling, was forced to help dig up the remains of other prisoners who'd been killed with them.

The once-secret reactor at HANFORD, WASHINGTON, which had played such an important role in creating the plutonium used in the Nagasaki and later bombs, contin-ued as a central site for the production of American nu-clear weapons. After several decades of service, though, a changed national mood increasingly saw it as a center of environmental despoliation: cleanup costs for its leaked or inadequately stored radiation were estimated at $30–$50 billion.

CECILIA PAYNE'S thesis advisor nearly brought her career to a halt by making sure she was kept from any of the new electronic equipment coming in. He also ensured, as direc-tor of Harvard's observatory, that when she did give courses, they weren't listed in the Harvard or Radcliffe cat-alog; she even found out, years later, that she had been clas-sified as "equipment expenses" when her salary came due. When the worst of the sexism ended, and a decent director of the observatory took over in the postwar era, it was too late. She had such a heavy teaching load by then that "there was literally no time for research, a setback from which I have never fully recovered."

Instead, she became one of the kindest supporters of the next generation at Radcliffe, always available for long talks to students at loose ends. She also kept intellectually nimble by learning languages, to add to the Latin, Greek, German, French, and Italian she'd been comfortable with when she'd arrived in America. "Icelandic was a minor challenge," her daughter wrote, though "I cannot say she truly mastered it." Cecilia Payne had the pleasure of seeing that daughter become an astronomer—and publishing sev-eral papers with her.

ARTHUR STANLEY EDDINGTON became increasingly resis-tant to the main trends of modern astronomy. One of his final works, published in 1939, had a chapter beginning "I believe there are 15 747 724 136 275 002 577 605 653 061 181 555 468 044 717 914 527 116 709 366 231 425 076 185 631 031 296 protons in the universe, and the same number of electrons." He was perplexed that professional astronomers stopped paying any attention to him.

In 1950, four years after FRED HOYLE'S paper on bomblike implosion inside stars, the merits of Cambridge nepotism were demonstrated when a director of radio talks from his old college overlooked the stern injunction against Hoyle in BBC files, and invited him anyway to give a series of broadcasts on astronomy. In the rush to prepare a script for the final talk, Hoyle coined a somewhat mocking phrase for a then-unproven theory about the origins of the universe. He called it the "Big Bang."

The BBC talks and subsequent book were such a success that not only did Hoyle and his wife get enough money to buy their first refrigerator, but it led to a career popularizing science, which he carried on in parallel with his academic research. This allowed him to put enough savings aside that in 1972, when he told Cambridge administrators he would resign if they continued going back on their word about funding for the successful astronomical research center he'd created, he was able to startle them ("Fred won't resign. Nobody resigns a Cambridge Chair"), and politely walk out. He has continued to publish innovative papers, some of them flighty, some of them profoundly sensible—as has been the wont of top scientists from Newton on. If it weren't for the way his Yorkshire honesty irritated the old guard in Britain and the astronomical community generally, it's generally accepted that he would have long since been granted a Nobel Prize for his work on the formation of the elements.

SUBRAHMANYAN CHANDRASEKHAR was renowned for keeping a calm exterior, but internally: "I am almost ashamed to confess it. Years run apace, but nothing done! I wish I had been more concentrated, directed and disciplined." At the time of this lament he was twenty, and it was but one year since the sea journey where he'd peered into the catch-22 from $E=mc^2$, which, along with other work, would ultimately lead to the understanding of black holes. He accepted a post at the University of Chicago, but his reserve meant that he and his wife settled in an observatory town over 100 miles from the main campus, largely so that they wouldn't have to embarrass Chicago faculty members by turning down invitations where alcohol or meat might be served. He diligently drove the full round-trip journey to Chicago for his teaching when needed, even during winter storms—once for a class that had only two students. (It was worth the drive, as that entire class—Yang and Lee—went on to win the Nobel Prize.)

Forty years after his rebuff by Eddington, Chandra finally returned to the study of black holes. There are photographs of brightly dressed young physicists in the clothes of the early 1970s, sitting around a table in the Caltech cafeteria, listening to this perfectly tailored, suited man of the generation of their grandparents. He surpassed almost all of them in his agility with new applications for general relativity, and in 1983, over half a century after the sea voyage, he published one of the fundamental works on the mathematical foundations of black holes. That was the year he won the Nobel Prize, and then—following his usual habit—he shifted directions once again, expanding an elaborate exploration of Shakespeare, and of esthetics generally.

In mid-1999, NASA launched a large satellite for deep space observation, capable of capturing images from the very edge of black holes. The satellite crosses over much of the earth—the Arabian Sea, Cambridge, and Chicago included—in its mission, and it is called the CHANDRA X-RAY OBSERVATORY.

Although ERWIN FREUNDLICH missed out on the 1919 eclipse expedition, his spirits recovered when industrialists in the new Weimar Republic donated large funds to build a great astronomical tower in Potsdam. This would allow further tests of general relativity's predictions, even in periods when there was no eclipse. Zeiss supplied the equipment, and Mendelsohn, the great expressionist architect, designed the building—it's the famous Einstein Tower featured in many books on 1920s German architecture.

Through Einstein's help, Freundlich became the Einstein Tower's scientific director. The measurements he undertook, however, proved to be impossible with the technology of the time. Only in 1960, at Harvard, did another team manage to give this further confirmation of Einstein's work.

$$E=mc^2$$

These notes are for people who want to know more. Some are serious: why Tom Stoppard has it all wrong when he uses relativity to try to back moral viewpoints in his plays; what the deep links are between relativity, thermodynamics, and the Talmud; how close Germany really came to getting radioactive weapons. Other notes are lighter, though also significant in their own way: I loved finding out that there are parts of World War I German battleships on the moon; that Maxwell didn't write Maxwell's equations, that Faraday never said, "Well, Prime Minister, someday you can tax it," and even why Einstein never liked calling his work the theory of relativity.

Preface

viii "Einstein explained his theory to me every day . . .": Carl Seelig, *Albert Einstein: A Documentary Biography* (London: Staples Press, 1956), pp. 80-81.

ix George Marshall saw to it . . . : Leslie Groves, *Now It Can Be Told: The Story of the Manhattan Project* (London: Andre Deutsch, 1963), pp. 199-201; Andrew Deutsch; see also Samuel Goudsmit, *Alsos: The Failure in German Science* (London: Sigma Books, 1947), p. 13.

1. Bern Patent Office, 1905

3   Letter to Professor Wilhelm Ostwald: *Collected Papers of Albert Einstein, Vol. I,* The Early Years: 1879-1902, trans. Anna Beck; consultant Peter Havas (Princeton, N.J.: Princeton University Press, 1987), p. 164. I've added Ostwald's address.

5   "... nothing would ever become of you ...": Ibid., p. xx.

5   "Your presence in the class ...": Philipp Frank, *Einstein: His Life and Times,* trans. George Rosen (New York: Knopf, 1947, revised 1953), p. 17.

5   "displayed some quite good achievements": Albrecht Fölsing, *Albert Einstein: A Biography* (London: Viking Penguin, 1997), pp. 115-16.

7   ... jokingly called his department of theoretical physics ...: The phrasing is recalled by a visitor, Rudolf Ladenburg; in Folsing, *Albert Einstein,* p. 222; see also Anton Reiser, *Albert Einstein, a Biographical Portrait* (New York: A. and C. Boni, 1930), p. 68.

7   "I like him a great deal ...": Folsing, *Albert Einstein,* p. 73.

7   ... feeling "the greatest excitement": Reiser, *Einstein,* p. 70.

7   "The idea is amusing ... that I cannot know." *Collected Papers,* vol. 5, doc. 28. The friend was Conrad Habicht.

8   $E=mc^2$ had arrived in the world: Einstein did not write $E=mc^2$ in 1905. In the symbols he was using at the time, the equation would have come out as $L=MV^2$. But more important, in 1905 he still only had the notion that when an object sends out energy, it will lose a small amount of mass in the process. The full understanding that the reverse happens only came later.

During World War II, when Einstein wrote out a copy of his main 1905 relativity paper to be auctioned for war bonds, he turned at one point to his secretary, Helen Dukas, as he was taking down her dictation: "Did I say that?" She told him he had. "I could have said it much more simply," he replied. (The story is in Banesh Hoffman, *Einstein, Creator and Rebel* (New York: Viking, 1972), p. 209.

2. E Is for Energy

11   One of the men who took a central role in changing this . . . : There were other researchers involved in understanding the conservation of energy, but focusing on Faraday gave me a chance to bring in the concept of a field pervading seemingly "empty" space, so central to Einstein's later work. For the other originators and their links, start with Thomas Kuhn's essay and Crosbie Smith's *The Science of Energy,* listed in the Guide to Further Reading for this chapter. Faraday's own views on how thoroughly energy was conserved differed from those of many subsequent researchers; see e.g., Joseph Agassi's *Faraday as Natural Philosopher* (Chicago: University of Chicago Press, 1971).

13   . . . a lecturer in Copenhagen had now found . . . : The Dane was Hans Christian Oersted, and most physics textbooks say that he "stumbled" across his results. But that's not possible: a compass needle won't be deflected if the compass is at an angle to the electric wire, or if the current in the wire is too low or too high, or if the wire is a low-resistance copper, and so on. In fact Oersted had been hunting for this link between electricity and magnetism for at least eight years. The reason that's so often missed is that his motivation hadn't come from standard scientists, but from Kant, Goethe (of the Elective Affinities), and, especially, Schelling. But Faraday recognized what Oersted had really been up to.

Note that Oersted's success doesn't mean all extrascientific motivations prove successful. An ability to objectively assess what such motivations offer is crucial. Einstein was excellent at this, at least early on in his career: his study of Hume readied him for seeing how arbitrary the woven definitions that physicists used were, and so how far they could someday be stretched; his love of Spinoza was a constant, urgent reminder of the ordered beauty waiting in our universe. Goethe, by contrast, was almost always poor at using philosophy in science, and wasted years on a theory of vision, simply because he was convinced it "should" be true. As the old saying has it, to do mathematics you need paper, a pen, and a wastebasket; to do philosophy, the paper and pen are enough.

14   "You know me as well or better . . .": The quote is from a letter from Faraday to Sarah Bernard, *The Correspondence of Michael Faraday,* vol. 1, ed. Frank A. J. L. James (London: Institute of Electrical Engineering, 1991), p. 199.

15   From his religious background, he imagined . . . : This is my own interpretation, building on ideas from cognitive anthropology on correlates between social behavior and ideologies. For a more conventional view, see Cantor's *Michael Faraday, Sandemanian and Scientist* in the Guide to Further Reading.

16   "Why, Prime Minister, someday you can tax it.": It's a catchy story, and pleasing for engineering types, but the phrase has never been found in Faraday's letters, or in the letters of anyone who knew him, or in any newspaper accounts of the time, or in any of the biographies written by individuals who had been close to him. American writers often recount it as having been said to Gladstone, which is less than convincing, as Gladstone became prime minister forty-seven years after Faraday's discovery, at a time when electrical devices were common. The British government had long been aware that its strength had grown with industrial innovation.

16   "All at once he exclaimed . . .": Silvanus P. Thompson, *Michael Faraday: His Life and Work* (London: Cassell, 1898), p. 51.

16   Faraday's invisible whirling lines . . . : It's the first modern occurrence of the notion of a "field." The reason this came as such a surprise in 1820s Europe was that for over a century, all respectable physicists had "known" that no such thing could exist. The medievals might have believed the heavens were full of goblins and spirits and unseeable, occult forces, but when Newton had shown how gravity could work instantaneously across empty space, without any intervening objects to carry it along, he had been "sweeping cobwebs off the sky."

Yet while others accepted that as given, Faraday researched enough to find that Newton himself had viewed the notion of entirely empty space as just a provisional step. Faraday liked quoting one of Newton's 1693 letters to the astronomically curious young theologian Bentley: ". . . that one body can act upon another at a distance, through a vacuum, without the mediation of anything else . . . is to me so great an absurdity, that I believe no man who has in philosophical matters a competent faculty of thinking can ever fall into it."

Both quotes are from Maxwell's essay "On Action at a Distance," in Volume II of *The Scientific Papers of James Clerk Maxwell*, ed. W. D. Niven. (Cambridge: Cambridge University Press, 1890), pp. 315, 316.

17   . . . and then Humphry Davy accused him . . . : What actually happened? It's true that Davy and the researcher William Hyde Wollaston had already started work on this topic, but Davy and Wollaston were nowhere near reaching Faraday's great result—and Faraday was not the sort to steal. To get an idea of Davy's hinted accusations, there are Faraday's distraught letters, and Wollaston's curt response, especially the letters of October 8 and November 1, 1821, in James, ed. *The Correspondence of Michael Faraday*. A measured discussion is available in *Michael Faraday: A Biography*, by L. Pearce Williams (London: Chapman and Hall, 1965), pp. 152-160.

18   Faraday never spoke out against Davy: But he was hurt. For years Faraday had been accumulating a scrapbook on Davy: there were geological sketches as a reminder of their travels together, drafts of several of Davy's papers, which Faraday had copied out in full in his own neat hand; friendly letters Davy had sent him in the past; little sketched doodles of events in their life. The scrapbook is arranged chronologically. After September 1821, Faraday never added to it again.

18   Letter of May 28, 1850, from Charles Dickens: *The Selected Correspondence of Michael Faraday*, vol. 2: 1849–1866, ed. L. Pearce Williams (Cambridge: Cambridge University Press), p. 583.

20–21   But Faraday's vision . . . a satisfactory alternative: In Faraday's time, energy conservation was just an empirical observation. Only in 1919 did Emmy Noether give a deeper explanation of why it was so persistently noted. For good introductions to the links between symmetry and conservation laws, see *The Force of Symmetry,* by Vincent Icke (Cambridge: Cambridge University Press, 1995), especially the discussion on p. 114; or Chapter 8 of *Fearful Symmetry: The Search for Beauty in Modern Physics,* by A. Zee (Princeton. N.J.: Princeton University Press, 1986).

21   this quiet school . . . student-centered lines: Einstein had the added advantage of lodging with Jost Winteler, the school's director, who twenty years before had completed an immensely original doctorate on the *Relativität der Verhältnisse,* or the "situational relativity" of surface features in a language, and how those stemmed from deeper, unchanging properties of the language's sound systems. The structural overlaps with Einstein's later work in physics are profound, down even to Einstein's preference for the label Invariant Theory for what he had created—the very term Winteler had used. For background on Winteler's thesis, see pp. 143ff of Roman Jakobson's contribution to *Albert Einstein, Historical and Cultural Perspectives,* ed. Gerald Holton and Yehuda Elkana (Princeton, N.J.: Princeton University Press, 1982); there's also the charming essay "My Favorite Topics" in *On Language: Roman Jakobson,* ed. Linda R. Waugh and Monique Monville-Burston (Cambridge, Mass.: Harvard University Press, 1990), pp. 61–66.

4. m Is for mass

28   [Lavoisier] . . . was the man who first showed . . . a single connected whole: The word *mass* is in quotes, for Lavoisier's findings were only about the conservation of matter, while in $E=mc^2$ the "m" stands for inertial *mass*. This is a much more general thing, concerned not with the detailed inner properties of an object, but simply, in the tradition of Galileo and Newton, with its overall resistance to being shifted or pushed. The distinction seems fussy, but is fundamental. Astronauts find themselves weighing less when they're on the moon than they did before leaving the Earth, but that isn't because parts of them have disappeared. In the same way, as we'll see in Chapter 5, if you

watch a sufficiently fast rocket, you'll see its mass increase immensely, but that happens without more atoms popping into being in its metal frame, or indeed without the atoms in its body getting pudgy at all.

What makes Lavoisier worth concentrating on is that his work on the conservation of matter ended up boosting interest in the conservation of mass, even though by today's understanding there's no reason for mass and matter to always be linked. In the late 1700s, however, no one cared that what he was "really" showing was the conservation of atoms—for no one in his time had a clear notion that atoms as physical entities existed.

29   the moment came . . . a truly major experiment: If one asks "Who was the first to show that the conservation of mass is true?" the answer has to be "No one, really." Lavoisier had shown in 1772 that some sort of air joined with metal when it was heated—but that was largely an extension of what de Morveau, Turgot, and others had done before him. In 1774 Lavoisier did carry out more extensive experiments with lead and tin, confirming that what carried the extra weight was air rushing into the heated containers—but this too wasn't entirely original, building on concepts he'd borrowed from the unsuspecting Englishman Priestley. Even the 1775 confirmatory experiments Lavoisier carried out with mercury, ended up being phrased in a way that atomists from Roman times would have taken for granted. Yet Lavoisier did more than grab credit for what others had done. Priestley and the others hadn't fully conceived of a conceptual system that made sense of these various experiments. Lavoisier had.

On the different attitudes brought to bear—as well as the historiographical considerations involved—see Simon Schaffer's accomplished "Measuring Virtue: Eudiometry, Enlightenment and Pneumatic Medicine," in *The Medical Enlightenment of the Eighteenth Century*, ed. A. Cunningham and R. K. French (Cambridge: Cambridge University Press, 1990), pp. 281–318.

33   "Everybody confirms that M. Lavoisier . . . French capital": Arthur Donovan, *Antoine Lavoisier: Science, Administration, and Revolution* (Oxford: Blackwell, 1993), p. 230.

34   "I am the anger, the just anger . . .": Louis Gottschalk, *Jean Paul Marat: A Study in Radicalism* (Chicago: University of Chicago Press, 1967).

34   "Our address is . . . room at the end": Letter from Lavoisier to his wife, November 30, 1793 (10 Frimaire, Year II); in Jean-Pierre Poirier, *Lavoisier: Chemist, Biologist, Economist* (College Park, Penn.: University of Pennsylvania Press, 1996), p. 356.

34   The trial itself was on May 8: It's common to read that when Lavoisier was condemned to death, the presiding judge declared, "The Revolution has no need for savants." But it's very unlikely that the president of the court, Jean-Baptiste Coffinhal, ever said that. The trial was not of individuals, but of the full group of senior members of the General Farm; Lavoisier was not singled out. Fairly detailed accounts of the proceedings survive. What infuriated the court and the jury—which included a barber, a stagecoach employee, a jeweler, and the former Marquis de Montfabert, now known simply as Dix-Août (August 10)—was the way the tax farmers had used their position to extort profits. Many scientists thrived during the Revolution, or at least survived by staying relatively quiet at the various interludes when passions were highest: Carnot, Monge, Laplace, Coulomb, and others. (The phrase "no need for savants" seems to have been invented two years later in a eulogy read by Antoine Fourcroy, a one-time student of Lavoisier's who'd become caught up in Revolutionary enthusiasm and now was trying to backpedal, showing that by all means he hadn't been cowardly in standing aside when his former mentor was attacked.)

35   "was led to the scaffold in a pitiful state": The witness was Eugène Cheverny, in Poirier, *Lavoisier,* p. 381.

35   Breathing was more of the same: With such insights, Lavoisier also became the founder of modern biology, by opening up the basics of physiology. Human blood, for example, is mostly water, and if you try to mix oxygen into water, not a great deal will stay there. But if you scatter some finely ground-up iron filings into the water, then oxygen you pump in will stick to that iron just as it did in

his lab. (Each iron fragment quickly starts rusting, and in doing so pulls in a great number of oxygen molecules, making them stick. The result is that the iron-rich water can hold on to a lot of oxygen.) This is how blood works: it's red for the same reason that the iron-rich clay soil of Georgia is red.

It was the promise of La Mettrie's *L'Homme Machine*; the same rush of optimism led to Lavoisier himself suggesting that in the future it would be possible to look inside the brain and see "the efforts required of someone giving a speech or . . . the mechanical energy spent by . . . a scholar who is writing, or a musician who is composing"— a pretty near description of our modern brain scans. The quote is from Lavoisier's *Collected Works,* vol. II, p. 697.

36   This is what Einstein was taught . . . different topics: The division of reality into two parts is something of a default operation of the human mind, seen in the ease with which we create the categories of friend or foe, right or wrong, x or not-x. The particular division bequeathed by Lavoisier, Faraday, and their colleagues was even more compelling, for when one of the divisions is material and physical, and the other is invisible yet still powerful, it's the ancient dichotomy of the body versus the soul that slips into our mind.

Many other thinkers have been guided in their work by that distinction. Alan Turing seems to have been led by the body-soul division when he came up with his distinction between software and hardware; most users of computers easily think that way, for we can all immediately grasp the notion of a "dead" physical substrate, powered up by a "live" controlling power. The soul-body distinction permeates our world: it's Don Quixote versus Sancho Panza; the cerebral Spock versus the stolid *Enterprise;* the contrast between the whispered encouraging voice-overs in the running-shoe ads, and the physical body on the screen.

But these lulling categories only make a suggestive division, not a proof. A young man such as Einstein, always keen to understand the foundation of a field for himself, could readily see that his professors had simply made an induction from a very incomplete data set.

There are many accounts of how lurking categories pull our thoughts along, as with George Lakoff and Mark Johnson's *Metaphors We Live By* (Chicago: University of Chicago Press, 1980), or Kedourie's excellent writings on nationalism, yet for some reason this author is especially pulled toward the approach in Bodanis's *Web of Words: The Ideas Behind Politics* (London: Macmillan, 1988).

38   . . . when members . . . in Florence: Galileo's proposal was in the First Day section of his *Two New Sciences*. The test was over twenty years later, probably around 1660, by the Accademia del Cimento in Florence. Their results are on page 158 of a book with the sort of vivid identifying location publishers no longer have: *Essayes of Natural Experiments, made in the Academie del Cimento*; Englished by Richard Waller, Fellow of the Royal Society, London. Printed for Benjamin Alsop at the Angel and Bible in the Poultrey, over-against the Church, 1634.

5. c Is for celeritas

40   The effort might be exhausting . . . embarrassing public exposure: Clearly I'm being slightly tongue in cheek about Cassini. From the available evidence, he might have been an insecure man, but as a newcomer to France he had a great deal to be insecure about: At first his appointment was only temporary, and he'd been warned not to try speaking French, but then he'd been told he *had* to learn French, for the Academy of Sciences couldn't be sullied by being exposed to Latin, let alone his native Italian. It's touching to read his own account of fearfully concentrating to try to develop the crucial language ability—and then his pride when complimented by the king on how much he'd progressed in just a few months. He also had a personal reason to resent Roemer. For Cassini himself had established his reputation by publicly proclaiming, back in July 1665, improved predictions for the transit of Jupiter's satellites. His predictions had been proven right in August and September of that year; his doubters had been humbled; the grand position in Paris had been his reward. He would not have appreciated Roemer trying to use the same twist against him.

There was also something more than mere pique in his critiques of Roemer for being so confident of the Jupiter observations. Cassini wrote a long poem, "Frammenti di Cosmografia," expressing his humility before the grandeur of space, and his belief that only an unjustified false pride lets humans, isolated on one inconsequential planet, presume they can accurately measure everything that occurs. Even before Roemer arrived, Cassini had applied first-order approximations to try to get rid of Io's anomalies; he was being sincere when he said it would be overhasty to insist on any one new interpretation. The poem and fragmentary autobiography are in *Mémoires pour Servir à l'Histoire des Sciences et à Celle de L'Observatoire Royal de Paris* (Paris, 1810), compiled by Cassini's great-grandson, also named Jean-Dominique; see especially pages 292 and 321.

45 once the underlying mathematics . . . could describe: Maxwell's later success has meant that other researchers of the time have tended to be overlooked. Weber at Göttingen is an especially interesting intermediate figure, for he too computed the speed of light in his efforts to link electricity and magnetism; but as it was masked by an extra factor of $\sqrt{2}$, he didn't recognize what he'd found, and left it aside. Weber's story is nicely described in M. Norton Wise's article "German Conceptions of Force . . . ," pp. 269–307 in *Conceptions of Ether: Studies in the History of Ether Theories 1740–1900,* ed. G. N. Cantor and M. J. S. Hodge (Cambridge: Cambridge University Press, 1981). Weber's caution is similar to that of the early Ampère; his hybrid equations—extending almost into the Maxwellian world of fields, but not quite making it—resemble a battleship forlornly loaded with antiaircraft guns.

45 "Aye, I suppose . . .": I fear this one is apocryphal too. All that's certain is that Maxwell liked making fun of his own literalness; also that he did experiment with staying up extremely late as a Cambridge student, to the astonishment and bemusement of his fellow students. See e.g., Goldman, *Demon in the Aether,* p. 62.

46 "They never understood me, but . . .": Ivan Tolstoy,

James Clerk Maxwell: A Biography (Edinburgh: Canongate, 1981), p. 20.

47 "As I proceeded with the study . . .": *Treatise on Electricity and Magnetism,* James Clerk Maxwell (Oxford: Clarendon Press, 1873); Maxwell's preface to his first edition, p. x.

47 When a light beam starts going forward . . . : Ordinary language is inherently inexact here, for what we're really describing are the properties of electrical and magnetic *fields,* a specification of what "would" happen at any given location. The subjunctive mood in grammar, and especially the subjunctive conditional, comes close to matching the idea: you might not be able to say what's happening at a particular street corner in a bad neighborhood right now, but you could tell what's liable to happen if a Rolex-laden tourist sauntered past. In the case of physics, think of the swirling curves of iron filings you can see around a bar magnet. Now take away those filings, and instead, where each one was, write down a number or a group of numbers that tell you how any filing placed there is likely to respond.

To someone who hadn't seen how you started, there would only be a cold list of numbers. But to someone who knew about the whirling power a magnet can have on iron filings, your list is a vivid description—and to Maxwell and Faraday, with their religious beliefs, your list would be a direct readout of the holy power that created that field in the first place.

47 The electricity and magnetism . . . "mutual embrace": It doesn't take much power to send out a wave. Play a piano key and the string simply vibrates back and forth, otherwise unmoving, while it's the *pattern* of those vibrations that moves along and carries the sound. There can be hundreds of gallons of air between two people standing a few yards apart in a corridor, yet they don't have to blow all that air forward to call out a hello. Each need only puff the tiny amount that can be shot up from their throats, and that will start a rippling compression wave that gets the job done.

Light and electromagnetic waves generally can be as

easy. Switch on the ignition of your car, and your spark plug sends out an electromagnetic wave that has several frequencies that will pass through the metal around it and make it streak to the orbit of the moon within two seconds of your hearing the engine catch; the wave continues outward and will have reached the distance of Jupiter a few hours later.

48 Maxwell's equations: Maxwell's work was a tremendous achievement—and would have been even more tremendous if he had ever written the four equations that bear his name. But he didn't. It's not just a matter of notation changing, for even the details Hertz later looked at, which led to the realization that radio waves could be transmitted and received like light waves, weren't in the equations Maxwell conceived.

The story of how Maxwell's equations were finally completed, by a group centered around three physicists in England and Ireland in the two decades after his death, is thoroughly recounted in Bruce J. Hunt's *The Maxwellians* (Ithaca, N.Y.: Cornell University Press, 1991).

50 . . . nothing can go faster: Or more precisely, nothing which started slower than the speed of light can end up going faster. What if there were particles—or perhaps an entire parallel world—located permanently on the "other side" of the light speed barrier? It sounds like science fiction, but physicists have learned to keep an open mind. (These postulated superluminal particles were labeled *tachyons* by Gerald Feinberg.) Another proviso is that we're discussing the speed of light in a vacuum, and light's velocity is lower in other substances. This is why diamonds sparkle: light skimming above the surface will go faster than the light that has dipped inside.

There are more significant exceptions—due to the effects of varying space-time curvatures on relative speeds; also there can be effects due to the role of negative energy, and there have also been intriguing results about pulses of light that exceed our velocity "c" (albeit in ways that keep added information from being transferred in the process). These, however, take us beyond the technical level of this book. I suspect future scientists will either look back at us in amazement that we ever took this seriously—or that we took so long to realize that this was the way to open up the first Disneyland in Andromeda.

52   . . . the solid mass of the shuttle starts to grow: None of our ordinary words apply well to this realm, and the term *swelling* has to be thought of as only a metaphor. The shuttle—or a proton, or any other object—doesn't get fatter in all directions. Rather, this is where the seemingly fussy distinction between conservation of matter and conservation of mass from the Lavoisier chapter comes into its own. With mass defined as the property of resistance to an acceleration—which is what we reflexively try to assess whenever we heft an object, to estimate its weight—then it's possible for mass to increase without matter swelling outward. So long as there is an increased resistance to an applied acceleration, the requirements are fulfilled.

In the slow speeds and ordinary realms we're used to, the amount of mass increase will not be enough to notice—this is why Einstein's predictions were so startling—but as an object moves away from us at rates approaching the speed of light, the effect gets clearer. The predictions are very precise.

The way to compute how much a given mass will increase is to take its velocity, square that, divide it by the square of the speed of light, subtract the result from 1, take the square root of *that,* then take the inverse, and then multiply that final result by the mass you're interested in. It's easier in shorthand: if a mass is traveling at the speed of "v," then to work out how much it will appear to swell, you multiply the original mass "m" by $1/\sqrt{(1-v^2/c^2)}$.

It usually helps in getting a feel for an equation to "tweak" it by examining its properties at various extreme values. If v is much less than c—i.e., if the shuttle is moving slowly—then $(1-v^2/c^2)$ is almost the same as one, for v^2/c^2 will be so slight. It doesn't matter if you then go on to take its square root and the inverse: you'll still get a number extremely close to 1. For the actual space shuttles launched from Florida, maximum v is about 18,000 mph. That's such a small percentage of the speed of light that its mass only expands by much less than a thousandth of a percent, even when it's screaming out of the atmosphere at top speed. But if a shuttle or anything else is moving really fast, so that v is close to c, then $(1-v^2/c^2)$ is close to zero. That means the square root is small also, and when you divide 1 by a small fraction, the result is huge. Watch an object streak past at 99 percent of the speed of light—and its mass will to you have increased several times.

There's a temptation to think this is just some quirk, and that although we might be confused in our measurements, a moving object won't "really" be more massive in any way that counts. But the magnets around the accelerator rings at CERN really do have to raise their power this much to keep a proton speeding at that rate on track, for otherwise the momentum of its increased mass will send it into the accelerator's walls. At 90 percent of the speed of light, the power needed to control 2.5 times more mass will have to be pumped in to avoid a skid out of the flight path and a crash. If the speed goes up to 99.9997 percent of the speed of light, then $1/\sqrt{(1-v^2/c^2)}$ gives a mass increase of 430 times—whence the problems of the accelerators at CERN, having to find some way of drawing their extra energy, without disturbing the good citizens of Geneva.

Simply to assert, however, that the expression $1/\sqrt{(1-v^2/c^2)}$ gives a rule you must follow, would put us in the same category as the obedient rule-following instructors Einstein had so resented. At davidbodanis.com we'll explore *why* it's true.

52   Energy that's pumped [in] . . . will turn into extra mass: The shuttle example is only heuristic; we'll see as the book goes on that energy *is* mass: the unified thing called "mass-energy" just happens to take on different aspects, depending on how we're viewing it. The restrictions of our fragile bodies means that we hardly ever change our speeds substantially, and so we view mass from a highly "skewed" angle. That distortion is the reason the "released" energy seems to be so high. (A significant proviso, however, is that this equivalence between mass and energy only holds when an object is viewed from the one particular view in which it's at rest. This is especially important in general relativity, for an object's gravitational attraction stems from its total

energy, and not just from its rest mass. Page 199 of the main text touches on this reasoning in connection with black holes; the point is developed at more length on our Web site.

6. ♎

56 "Supervising the workers": *Voltaire et la Societé Francaise au XVIII è Siècle: Volume I, La Jeunesse de Voltaire,* by Gustave Desnoiresterres (Paris: Dider et Cie, 1867), p. 345.

56 . . . a new concept in the air . . . : Arouet didn't need Newton's work to make him aware of France's faults. If anything, it wasn't abstract ideas, but seeing England's working parliament—and the tradition of at least semi-independent judges and citizens' rights—which helped in showing up the lackings in France. But it was sweet to have the backing of the world's most-renowned analytic system in these critiques. See Voltaire's *English Letters.*

56 Newton had created a system of laws . . . : Curiously enough, seeing an apple fall does seem to have helped Newton take the final step. William Stukeley recorded the elderly Newton reminiscing in his account published two centuries later as *Memoirs of Sir Isaac Newton's Life* (London: Taylor & Francis, 1936), pp. 19-20.

> *After dinner, the weather being warm, we went into the garden [of Newton's last residence, in London's Kensington] and drank thea, under the shade of some appletrees, only he and myself. Amidst other discourse, he told me, he was just in the same situation, as when formerly, the notion of gravitation came into his mind. It was occasion'd by the fall of an apple, as he sat in a contemplative mood. Why should that apple always descend . . . constantly to the earths centre? Assuredly, the reason is, that . . . there must be a drawing power in matter . . . like that we here call gravity, which extends its self thro' the universe.*

That was how Newton could be so sure that the forces on Earth are the same as those operating up in space. It's easy enough to measure the speed at which an object on Earth falls. In a single second, a dropped apple—or any other object—will fall about 16 feet. But how to measure the speed at which the moon "falls" to compare with that?

The way to do this is to recognize that the moon constantly falls downward, at least a little bit. (If it didn't fall, and only moved in a perfectly straight line, then it would soon shoot away from our planet.) The amount that it "falls" is just enough to keep it curving around the earth. Knowing the length of its orbit, and the amount of time it takes to make one circuit, one can conclude that the moon is tumbling earthward at about 1/20 inch every second.

At first that seems like a failure of Newton's guess. If there's some force making rocks fall 16 feet in one second down on Earth, then one might think that only a very different sort of force, out in distant space, would make giant rocks such as the moon fall a scant 1/20 inch in every second. Even taking into account the greater distance of the moon, it doesn't seem to work. The Earth is about 8,000 miles thick, so Newton, as well as his mother's apple trees, existed about 4,000 miles above the center of the Earth. The moon is in orbit about 240,000 miles from the center of the Earth, i.e., about 60 times farther. Even if you weakened a rock's fall by 60 times, it would still not flutter downward as slowly as the moon. (1/60 of 16 feet is about 3 inches—still far more than the scant 1/20th inch the moon falls each second.)

But what if you imagine a force that weakened by 60 *times* 60 times as it stretched up and away from our planet? It's an interesting idea—that gravity acts in accord with the square of the distance between objects—but how could you verify such a thing? You would have to prove somehow that gravity produces a force 3,600 times (60 x 60) stronger on Earth than out in space. No one in the seventeenth century—not even from Cambridge—could rocket up to the moon and compare the force of gravity there with what it is on Earth. But no one needed to. The power of equations is immense. Newton had the answer all along. "Why should that apple," he'd asked, "always descend . . . constantly to the earths centre?" In one second on the Earth's surface, a rock or an apple or even an astonished Cambridge don will fall 16 feet. But the moon in that time will

NOTES

fall just 1/20th inch. Divide the two numbers, and you have the ratio: how much stronger gravity's tugging power is on the Earth's surface than up on the moon.

It's just about 3,600 times.

That was the calculation Newton did, pretty much, in 1666. Imagine a giant clock, where the moon and earth were parts. Newton's rule showed, exactly, how the invisible connecting cogs and rods held the whole swirling contraption together. Anyone reading Newton, and following this argument, could gaze up and understand, for the first time, that the tug of gravity on their body was the same force that reached up, stretching on to the orbit of the moon and forever beyond.

58 "My youngest flaunts her mind . . .": Samuel Edwards, *The Divine Mistress* (London: Cassell, 1971), p. 12.

58 . . . memorize cards at the blackjack table: But even this, in her family's opinion, was something she got wrong. "My daughter is mad," her father wrote in exasperation. "Last week she won more than two thousand gold louis at the card tables, and after ordering new gowns . . . she spent the other half on new books. . . . She would not understand that no great lord will marry a woman who is seen reading every day." Ibid., p. 11.

59 "I was tired of the lazy, quarrelsome life . . .": Voltaire's *Mémoires;* in Edwards, *The Divine Mistress,* p. 85.

60 "is changing staircases into chimneys . . .": Letter from Voltaire to Mme de la Neuville, in André Maurel, *The Romance of Mme du Châtelet and Voltaire,* trans. Walter Mastyn (London: Hatchette, 1930).

61 . . . he discovered her with another lover . . . putting him at ease . . . : The various accounts—by servants as well as participants—of this incident are compared in René Vaillot's *Voltaire en son temps: avec Mme du Châtelet 1734–1748,* published in French by the Voltaire Foundation, Taylor Institution, Oxford England 1988.

61 The occasional visitors from Versailles . . . : The most thorough description is from Mme de Graffigny's *Vie privée de Voltaire et de Mme de Châtelet* (Paris, 1820).

62 She knew that most people felt energy . . . : The word *energy* is anachronistic here, for we're describing the period when these concepts were still being formed. But I think it captures the underlying ideas of the time. See, e.g., L. Laudan, "The *vis visa* controversy, a post mortem," *Isis, 59* (1968), pp. 131–43.

63 Along with various abstract geometric arguments . . . : Galileo had found that freely tumbling objects don't fall at an unchanging rate. Instead of covering a fixed amount of distance each second, they'll cover 1 unit of distance in the first second, 3 units in the second, 5 units in the third, and so on. Add that sequence of odd numbers together, and you get the *accumulated* distance a falling object travels: In the first second it's 1 unit, in the second it's 4 units (1 + 3), in the third second it's 9 units (1 + 3 + 5), etc. Through a mix of theory and experiment, this was the basis of Galileo's famous result, that accumulated distance is proportional to the square of the amount of time an object's been falling, or $d \propto t^2$. Leibniz extended this reasoning.

63 "According to [Newton's] doctrine . . .": Richard Westfall, *Never at Rest: A Biography of Isaac Newton* (Cambridge: Cambridge University Press, 1987), pp. 777-78.

65 . . . for du Châtelet it was one of the peak moments of her life . . . : The issue is more complex than either Newton or Leibniz recognized, and it took the impartial du Châtelet to understand what was valid and had to be preserved in both. Newton really did have a good point, despite Leibniz's mocking, for if the stars were spread randomly, why shouldn't gravity simply make them fall towards each other? And Leibniz also had a good point, for he never asserted that there was a perfect interventionist God, but merely that there was an optimal deity, subject to constraints we might not be able to see. This was a very different matter. Voltaire missed the point in his powerful satire *Candide*, but it became a fundamental principle in physics. In a variant form, it became central to Einstein's general relativity, where—as we'll see in the epilogue—planets and stars move in optimal paths within the curved spacetime of the universe.

What effect did it have on Voltaire to see du Châtelet puzzling through these issues? He would constantly be reminded of the contrast between the vast universe and the little "atom of mud" on which vain humans existed—which was one central theme in his work. He would also constantly be reminded of the need to give space for individual genius—a theme that life with the exhausting, exhilarating du Châtelet would no doubt tend to reinforce.

65 Willem 'sGravesande: The last name is not a misprint; the symbol *'s* means "of the" and is still common in Dutch: The city Den Haag (The Hague) is officially called 's-gravenhage (the hague of the Earls). I'm simplifying a large range of experiments 'sGravesande carried out: He used bullet-shaped ivory cylinders, hollow and solid brass balls, pendulums, scraped clay (of deeply elaborate consistency), supporting frames, and a Laputian-like variety of other contraptions to carry out his contention that "The Properties of Body cannot be known *à priori;* we must therefore examine Body itself, and nicely consider all its Properties . . ." See his (most beautifully illustrated) *Mathematical Elements of Natural Philosophy, Confirm'd by Experiments,* trans. J. T. Desaguilliers, especially Book II, ch. 3, 6th edition (London: 1747); the quote is from p. iv.

66 "In this, our delightful retreat": Voltaire's *Mémoires,* in Edwards, *The Divine Mistress,* p. 86.

66 "I am pregnant . . .": Letter to Mme de Boufflers, April 3, 1749. In *Les lettres de la Marquise du Châtelet,* vol. 2, ed. T. Besterman (Geneva: Institut et Musee Voltaire, 1958), p. 247.

67 "I have lost the half of myself . . .": Voltaire to d'Argental, in, e.g., Frank Hamel, *An Eighteenth-Century Marquise* (London: Stanley Paul & Co., 1910), p. 369.

68 A car that's racing along at four times another one's speed . . . : A wind of 20 mph is gentle, but a wind of 200 mph is catastrophic, and more like an ignited gas stove exploding. It's a lot more than 10 times as powerful, for it carries 10^2 or *100* times more energy. That's also why jetliners have to travel so high. Only the thinner air up there lets the

plane survive hours in the immense power of a 600-mph storm that the plane's speed produces.

Athletes perform these complex calculations all the time. Most schoolchildren can toss a ball at 20 mph, but only a few professional athletes can throw a ball at 100 mph. It's "only" five times as fast, but since energy goes up as the *square* of the speed ($E = mv^2$), the athlete has to generate 25 times as much energy. What's more, she has to do it in only $1/5$ the time. (For if the athlete took exactly as long to move her arm as the child did, the ball would come out at only 20 mph.) To pour out 25 times more energy, in $1/5$ the time, means she needs to generate 25×5 or 125 times more power! Other effects such as air resistance make it even harder. The one factor that does help an adult athlete is having a longer lever arm than a child.

68   Only by concentrating on mv^2: The point is not that mv^2 is "true," while mv^1 is not. Newton's concept of momentum—mv^1—is quite central to our understanding of the universe. Rather, each definition carves out different domains—different aspects—for us to concentrate on. Shoot a rifle, and the recoil is best understood in terms of mv^1; the impact of the bullet in terms of mv^2. A rifle and a bullet will have equal amounts of momentum right after the trigger is pulled, but the kickback fom a moving gun won't kill you: most of its kinetic energy is carried in its mass, so its velocity will have very little effect on the shooter. The bullet fired out however has so little mass that it carries most of the same momentum in its velocity. It's the square of that velocity—the bullet's kinetic energy—which signals how dangerous its high-speed flight can be to its target.

68   This isn't a proof, of course . . . : This is one of the items developed on my Web site.

69   That enormous conversion factor . . . of the equation.: If mass were too easily converted fully to energy, pens and pencils around us would start going off with blinding flashes of light, taking much of the Earth's cities with them; most of the physical universe would soon blast out of material existence.

What saves us is the principle of baryon conservation, which holds, roughly, that the total of protons and neutrons in the universe does not change; they cannot abruptly start disappearing.

The one time when 100 percent conversion does occur is when ordinary matter bumps against antimatter. A typical proton in our bodies has a baryon number of $+1$, but an antimatter antiproton has a baryon number of -1, so if they ever did annihilate, the sum of baryons in the universe wouldn't have changed. We actually experience something related to this every day, for a portion of the radon gas that wafts up from basements or out of walls produces antimatter as part of its decay process. Where that contacts ordinary air molecules or our skin, a (small-scale!) explosion from $E=mc^2$ operating at full power immediately results.

7. Einstein and the Equation

74   . . . rocking his one-year-old . . . : From a slightly later period, see e.g., D. Reichinstein's reminiscences, collected in his *Albert Einstein: A Picture of His Life and His Conception of The World*, by David Reichinstein (London: Edward Goldston, Ltd, 1934).

74   . . . as he referred to his notion of God . . . : For a particularly thoughtful analysis, see Max Jammer's *Einstein and Religion: Physics and Theology* (Princeton, N.J.: Princeton University Press, 1999). A rich compilation of current views by scientists about religion—both pro and con—is in Russell Stannard's *Science and Wonders: Conversations about Science and Belief* (London: Faber and Faber, 1996), which developed from a BBC radio series.

75   "We are in the position . . .": Einstein goes on: "That, it seems to me, is the attitude of the human mind, even the greatest and most cultured, toward God." From a 1929 interview with the then-famous journalist George Sylvester Viereck, reported in Viereck's *Glimpses of the Great* (London: Duckworth, 1934), p. 372. The wording is probably only approximate, as Viereck at other points admits finding his own shorthand notes hard to decipher.

75   In fact she had a passionate nature . . . : The affair is described in *Marie Curie: A Life,* by Susan Quinn (orig. New York: Simon & Schuster, 1995; London: Mandarin pbk., 1996). Einstein's quote stems from a 1913 hike together, reported on p. 348 of the UK edition.

78   He sent off the relativity article . . . : The epoch-making article was rejected for several reasons, not least of which was the sound bureaucratic grounds that it was a printed document and ". . . the regulations insisted upon a hand-written thesis." Carl Seelig, *Albert Einstein: A Documentary Biography* (London: Staples Press, 1956), p. 88 has the story from Paul Gruner, a supporter of Einstein's, who witnessed this among the Bern faculty.

79   ". . . like a man whose skin has been peeled off . . . ": Marianne Weber, *Max Weber: A Biography,* ed. and trans. Harry Zohn (New York: John Wiley & Sons, 1975), p. 286.

80   . . . time advanced smoothly . . . : Einstein wasn't the first to see that Newton's laws were consistent with there being no external "authority" or measurement standard by which our particular activities could be judged—for Newton had seen it as well! But in an intensely theological era, Newton had to be discreet in his thoughts while he was venturing into heresy this way. It was to a great extent to avoid such a God-denying "free float" of time that Newton incorporated absolute time in his *Principia.* The standard account is in Newton's General Scholium, but do see his more colloquial sequence of explanations in the letters to Richard Bentley (then a young theologian); both are conveniently available in the Norton reader *Newton: Texts, Backgrounds, Commentaries,* ed. Bernard Cohen and Richard Westfall (New York: Norton, 1995). Would Newton have taken the few simple algebraic steps to develop special relativity if he hadn't been held back by these cautions?

80   . . . a world where . . . an easy 30 mph: The image stems from George Gamow's estimable Mr. Tompkins series, on which generations of science lovers were raised. When Gamow wrote it the vision was a great fantasy; I think he would have been pleased that before the twentieth century was over "in February, 1999" a team at Harvard used laser-

cooling to produce a substance in such a state "it was within 50 billionths of a degree of absolute zero" that light was measured by outside observers to travel at just under 39 mph in it.

81   . . . come down to their original, static weight: The ordinary terms such as *weight* or *bulk up* in mass are, again, just to give an indication of what's going on.

81   . . . passengers inside would appear to have shrunk . . . : Textbooks usually say that a speeding car's length will be contracted, shrinking to the thickness of tissue paper. But although a direct application of the contraction factor from the note on page 249 would suggest this to be the case, what actually occurs is more subtle, due to such effects as the way light coming from different parts of the car must be emitted at different times. Distortions are similar to the way the three-dimensional Earth gets skewed when it's converted into two-dimensional Mercator projections for maps.

83   The Global Positioning System . . . : Along with the corrections applied to GPS satellite signals that do stem from special relativity, substantial effects are also due to general relativistic considerations, as is well surveyed in Clifford M. Will, *Was Einstein Right: Putting General Relativity to the Test* (Oxford: Oxford University Press, 1993). I love the idea that millions of people, holding GPS receivers at one time or another, have wrapped their hands around devices that contain miniaturized transpositions of the logical sequences that once occurred in Einstein's brain.

84   . . . the label *relativity* . . . : Einstein never used the phrase *theory of relativity* in his original 1905 paper; this was only suggested by Planck and others a year later. The name he really liked came from Minkowski, in 1908, who referred, accurately, to Einstein's "Invariant Postulates." If that had taken, we'd talk about Albert Einstein and his famous "theory of invariants." But by the time there was a wider move to make such a change, in the 1920s, the original, unwanted label had stuck.

84   . . . the wrong impression . . . : "The meaning of rela-

tivity has been widely misunderstood," Einstein explained in 1929. "Philosophers play with the word, like a child with a doll . . . It [relativity] does not mean that everything in life is relative."

Einstein was misinterpreted, in large part, because many people were *ready* to misinterpret him. Cézanne had spoken about the need to focus only on what you, personally, see and measure: a patch of red here, a glob of blue there. This was taken to match the way relativity questioned there being an impersonal, "objective" background world, just waiting, like a given interpretation of a Paris boulevard, for everyone to share. More recently, Tom Stoppard—who likes to undercut conventional perspectives—is happy to have characters in his plays refer to Einsteinian effects that seem to back that view.

The problem, though, is that these uses have nothing to do with Einstein's work. As mentioned in the main text, the divergence from any usual effects is far too small to be noticed at the speeds where we normally live. Even more central, the fact that the theory actually hinges on a few key invariants being preserved—the speed of light; the consistency and "equality" of any given coordinate frame—is quite the *opposite* of how the theory is commonly presented. Einstein himself once explained this to an art historian who had been trying to link Cubism with the theory of relativity:

> The essence of the theory of relativity has been incorrectly understood. . . . The theory says only that . . . general laws are such that their form does not depend on the choice of the system of coordinates. This logical demand, however, has nothing to do with how the single, specific case is represented. A multiplicity of systems of coordinates is not [emphasis added] needed for its representation. It is completely sufficient to describe the whole mathematically in relation to one system of coordinates.
>
> This is quite different in the case of Picasso's painting . . . This new artistic "language" has nothing in common with the Theory of Relativity.

The quote is in Paul LaPorte, "Cubism and Relativity, with a Letter of Albert Einstein," *Art Journal*, 25, no. 3 (1966),

p. 246; quoted in Gerald Holton's *The Advancement of Science, and Its Burdens* (Cambridge, Mass.: Harvard University Press, 1998), p. 109. Holton goes on to make the sensible observation that the notion of a multiplicity of possible observational frames is, in fact, the essence of all modern science, where modern means anything from Galileo in the early 1600s on; multiple and consistent views of a drawing have been a commonplace of architects for a very long time as well.

84   ... both Einstein and Newton ... impossibly brief periods ... : Here the laurel has to go to Einstein. Newton is famous for stating that he discovered differential calculus, the composition of light, and universal gravitation, all in the brief period when he was at his mother's farm. But when he recalled this he was already a very old man, talking up the past. The calculations he'd made on the farm had not been very persuasive—instead of the figure of 3,600 we used for the weakness of Earth's gravity at the orbit of the moon, which would "prove" an inverse square law of gravity applied, inaccuracies in measurements of the Earth meant that the best figure he actually got was a far from convincing 4,300 or so. He was also confused about the role of centrifugal force, and whether or not the moon was spinning in a Descartes-style vortex—there was a great deal of work to do when he got back to Cambridge. But then, humility is probably not what's needed for anyone working at these levels.

The fluently written article "Newton and the Eötvös Experiment" by Curtis Wilson, in his collected works *Astronomy from Kepler to Newton: Historical Studies,* (London: Variorum Reprints, 1989), is especially good on the subtleties Newton still had to work through. Chapter five of Westfall's *Never at Rest: A Biography of Isaac Newton* (Cambridge: Cambridge University Press, 1987) further examines Newton's actual accomplishments during the plague years in Lincolnshire; see also *The "Annus Mirabilis" of Sir Isaac Newton, 1666-1966,* ed. Robert Palter (Cambridge, Mass.: MIT Press, 1970).

86   They would learn what was on offer ... : What I like about Veblen is that he concentrates on a particular so-

ciointellectual cusp—the intersection of religion with science—which is liable to be especially laden with meaning. We can go deeper into Einstein's own work to see that in operation.

The first thing that leaps out is Einstein's great belief in unity. One part of traditional physics was built up from conventional Newtonian mechanics, where there were always ways to compare two different observers: to see who was going faster or slower than the other; to determine, objectively, that someone who was speeding in a car and switched a headlight on would make that light beam move along "faster" than would someone in a car that was standing still. But on the other hand, as Einstein realized, the other part of traditional physics built on Maxwell's development of Faraday's work, and that hinged on the speed of light appearing the same from the vantage point of any smoothly moving observer. Both a stationary driver and a moving driver would have to see any headlight beam shoot forward at 670 million mph. To Newton that was impossible. To Maxwell it was indispensable.

Most of the other researchers who bothered with this had just shrugged it aside, but to Einstein (in a 1920 paper quoted in Fölsing, p. 171), "The idea that these were two disparate situations was intolerable to me." For Einstein often said that one of his deepest ethical/religious beliefs was the ideal of social justice. What seems an unjustified or unfair distinction should, if one only examines it closely enough, be able to be resolved so that the unfairness no longer exists. It's the fairness principle of John Rawls, and all other believers in the objectionability of undeserved distinctions; a transposition outward, of the unified dominion, which a unitary deity might be expected to create.

What Einstein did to resolve this dilemma of how Newton "contradicted" Maxwell was to take one of those right-angle jumps of the sort Faraday and Roemer had been so successful with before. He questioned the very terms in which the dilemma was posed! The definitions of length and time and simultaneity had been around so long—they'd been codified at least since Newton—that they seemed "basic" to common sense. But Einstein realized they all contained loaded assumptions about how measurements had to be made. Newton and Maxwell were being puuuuuuuulled apart . . . and Einstein let a change in the way of building up the definitions that both had been assuming snap to take up the slack.

If I say a light beam should have passed a certain target pole by now but you say that's crazy, it's definitely going to take longer, that's no problem at all—*so long as your notion of longer is different from my notion of what that longer must be.* What I see is then true, *and* it's not conflicting with what you see. Throughout special relativity, what had seemed contradictions are resolved by clarifying our perceptual terms.

Was it a revolution? Einstein always insisted it wasn't, and that by shifting the core notions he was simply doing what was necessary to preserve the past. I'd take him at his word. Possibly his drive for continuity—for preserving the essence of a past—was at heart a desire for religious continuity; possibly it was his own respect for the great physicists of the past.

I suspect it also was reinforced by all his years of traveling. First there had been the soft Swabian home, then a harsh Prussian-style school, in the setting of Catholic Bavaria; next came a few delightful teenage months in the freer air of Italy, followed by the intense mix of intellectual and romantic attachments in distant, isolated Aarau; after that, life in student Zurich, with the disappointments of a narrow, cold Polytechnic faculty in the background; then quickly, Bern and the rush of adult obligation—of wife, and children, and obeisance within a great civil service hierarchy. At that point Einstein was still only in his early 20s: Lorentz at that age had never been out of the Netherlands. Later, for Einstein, there would be more countries, more cities; ending only at Princeton, in a distant, barely understood America. In such traveling, such isolation, the one thing that travels intact is yourself.

86   . . . different views about personal responsibility . . . : What type of theory is the theory of relativity that Einstein created? It's not like detailed laws such as the ones found in engineering texts, which might say that air resistance goes up as a certain power of an airplane's speed. Looked

at in greater detail, such "laws" will break down, for their assumptions are based on only partial analyses. They're more just useful rules of thumb, carved out to conveniently summarize subsets of the physical world we're especially concerned with, but which go no further than that.

Other principles, such as Newton's Third Law of Motion—the one about every action having an equal and opposite reaction—go deeper. They're what would be used in improving engineering rules of thumb about air resistance, for they're much more deeply embedded in the nature of an analytic system. Their application in such systems is, in principle, unlimited.

Einstein's special theory of relativity is different yet again. It's not a particular result, which simply happens to go beyond Newton or Maxwell's work. Rather, it's a theory *about* theories: the specification of the two criteria—that the speed of light is the same to all observers; that no smoothly moving reference frame is inherently indistinguishable from any other—which any valid theory must fulfill. If those criteria hold, the theory being considered might be true. If not, then it is definitely false.

Special relativity is simply a judgment machine. It's a meta-level commentary, just like the layered analyses of the Talmud; just like the Second Law of Thermodynamics.

This juridical nature of Einstein's theory is often missed, for after enunciating his principle, Einstein himself—and then many others—went on to come up with particular results, such as $E=mc^2$ or the observed slowing of time, which seem analogous to the derived particularities other theories generate. Yet this higher-order nature of the law is why the "m" in $E=mc^2$ is so general, applying to every substance in the universe—from the carbon in your hand, to plutonium in a bomb, or the hydrogen inside the sun.

87   "a temptation to superficiality . . . : Albrecht Fölsing, *Albert Einstein: A Biography* (London: Viking Penguin, 1997), p. 102.

87   . . . this gently self-teasing tone: In the letters of many artists of the time there's often a similar tone; a similarly bemused acceptance that there is a less than rational world

of received rules within which we have to live. The fact that an entire knowledge-admiring academic system was mixed in with a society that had totally different standards—of Junker superiority; of Kaiserlich grandeur—roused intelligent cynicism among many of the young.

87 "This should suffice to show . . . Albert Einstein never did attain.": The quotes are from his sister Maja's delightful short memoir, "Albert Einstein—A Biographical Sketch," in *Collected Papers of Albert Einstein, Vol. 1. The Early Years: 1879–1902,* trans. Anna Beck, consultant Peter Havas (Princeton, N.J.: Princeton University Press, 1987).

88 Uncle Rudolf ("The Rich"): Ibid., p. 160.

88 "he is oppressed by the thought . . .": Ibid., p. 164. It's Hermann Einstein's 1901 letter to Professor Ostwald again.

88 Eventually a few other physicists . . . : Planck's student Max von Laue was the first to see the great professor who had written this paper. Von Laue was directed from the patent office reception room and along a corridor; a young man came out whom von Laue ignored; von Laue waited. Later the young man returned. It was Einstein; finally the two said hello. From von Laue's account in a 1952 letter in Carl Seelig, *Albert Einstein: A Documentary Biography* (trans. Mervyn Savill (London: Staples Press, 1956), p. 78.

89 "I have to tell you . . .": *Collected Papers of Albert Einstein,* vol. 1. I've rearranged material from the letters given in documents 39, 72, 76, and 70.

90 The myth that she had been responsible for . . . : The story was first promoted in *In the Shadow of Albert Einstein*, published in Serbo-Croat in 1969 by the retired schoolteacher Desanka Trbuhović-Gjurić. It was developed in Andrea Gabor's *Einstein's Wife* (New York: Viking, 1995), and received a great public boost when Jill Ker Conway, onetime Smith College president, reviewed Gabor's book most favorably in *The New York Times*.

Alas, the *Times* and Conway (and Gabor, and Trbuhović-Gjurić) had the story totally wrong. Mileva was a good physics undergraduate, but no muse. See John Stachel's "Albert Einstein and Mileva Marić: A Collabora-

tion that Failed to Develop" in *Creative Couples in Science,*
ed. H. Pycior, N. Slack, and P. Abir-Am (New Brunswick,
N.J.: Rutgers University Press, 1995). The real quality of
their relation is best seen in *Albert Einstein, Mileva Marić: The
Love Letters,* ed. Jurgen Renn and Robert Schulmann; trans.
Shawn Smith (Princeton, N.J.: Princeton University Press,
1992).

91 Compared with this . . . child's play: Banesh Hoffmann, *Albert Einstein: Creator and Rebel* (New York: Viking,
1972), p. 116.

8. Into the Atom

94 Their finding is so widely taught . . . : Rutherford first
suspected that each atom would be a diffuse blob of electricity—since many physicists had gone to English schools,
images of raisin puddings were common in the textbooks
of the time. But when he created something like an ultra-
miniature atomic bazooka and started shooting alpha particles into gold foil, a few ricocheted back and he knew
something solid was tucked away in there. But where?

This was a great problem, for although Rutherford was
one of the century's finest experimentalists, he was something of an embarrassment as a mathematician. He couldn't
work out a plausible trajectory for what was happening to
the particles that he'd shot in and which had then somehow curved around and roared back out. As a result, he
ended up borrowing the mathematics of conic sections,
which had been developed in classical times and used in
the seventeenth century to track comet orbits. It worked to
some extent—in time he did manage to get his Manchester
results to seem to fit—but it also meant that for years students were taught that the atom really *was* like a miniature
solar system. That doesn't make sense, however: there's no
reason the electrons wouldn't crash inward after emitting
radiation in their fast orbits; nor is there any physical ana-
logue to the stability of the actual solar system, guaranteed
by Newtonian inverse-square gravity. But such is the power
of assumption-loaded mathematics (and also, who had a
better idea?) that although the solar system model was

eventually overthrown, what began with the mathematical weakness of Ernest Rutherford has carried on in popular mythology to become the default model most people carry of what an atom looks like.

96   There were positively charged particles . . . : How could one tell there was a positive charge in the nucleus? The reason is the old law from high school classes: similar electric charges repel, and opposite charges attract. If you shot a positive particle into the center and it stuck, you'd guess there was a negative charge waiting there. But the incoming alpha particles Rutherford shot in were positively charged and deflected away from "something" hovering at the center of atoms. That "something" had to be positive as well.

96   . . . the "Fewtron": Andrew Brown, *The Neutron and the Bomb: A Biography of Sir James Chadwick* (Oxford: Oxford University Press, 1997), p. 103.

98   The reason they [the slowed neutrons] stuck so well . . . : The reference to quantum uncertainty is further discussed in the notes to Chapter 10.

98   . . . his assistants lugged up buckets of water . . . : What worked in Fermi's research villa could work anywhere lots of neutron-slowing water gushed around radioactive clumps. In the early 1970s mining geologists were puzzled by peculiar ore specimens from a mine near the Oklo river in Gabon. Specialists from the French Atomic Energy Commission soon realized that this was where a series of natural uranium deposits had gone critical, over 1.8 billion years earlier. A natural aquifer had supplied the needed water; each of the reactions had continued for up to 100,000 years before dying down.

98   George de Hevesy employed it . . . : De Hevesy's culinary defense was undertaken with lead and similar elements two decades before Fermi's work. See M. A. Tuve's "The New Alchemy," *Radiology*, vol. 35 (Aug. 1940), p. 180.

9. Quiet in the Midday Snow

100   "I have here . . . no position . . .": Sallie Watkins's essay "Lise Meitner: The Foiled Nobelist", in Rayner-

Canham, *A Devotion to Their Science* (Toronto: McGill—Queen's University Press, 1997), p. 184.

100 "Our Madame Curie": Philipp Frank, *Einstein: His Life and Times* (New York: Knopf, 1947), pp. 111-12.

101 "[Hahn] would whistle large sections . . .": Ruth Lewin Sime, *Lise Meitner: A Life in Physics* (Berkeley: University of California Press, 1996), p. 35.

102 "I've fallen in love . . .": Sime, *Lise Meitner,* p. 37.

103 "Dear Herr Hahn! . . .": Ibid., pp. 69 and 67. The extracts are from letters of January 17, 1918, and August 6, 1917.

103 . . . Meitner shifted once again . . . : There was a certain amount of envy from less successful Berlin peers when Meitner began concentrating on the neutron, and a certain amount of disgruntled gossip as well. It's rare to change a lab's research direction: all the equipment is set up for one sort of work; there are postgrads whose grants are contingent on that previous work, technicians who were trained for it, and sometimes even suppliers who've come to specialize in it. Economists call it the problem of sunk costs, and it's one of the main reasons that very few top labs stay at the top for long. In a more recent era, it's why computer industry monoliths have continually been wrong-footed by quick Silicon Valley startups. Despite her surface shyness, Meitner would have been a confident dot-com entrepreneur par excellence.

104 "The Jewess endangers our institute . . .": For the Kurt Hess quote and associated details: Sallie Watkins's essay in *A Devotion to Their Science,* p. 183; also Sime, *Lise Meitner,* pp. 184-85.

104 Hahn may have been slightly troubled . . . : There are many levels of culpability, and Hahn of course was never a Nazi. Indeed, several months after Hitler came to power, Hahn had suggested to Planck that there should be a protest against the way Jewish academics were being expelled. By the late 1930s, such public protests were impossible, but a number of other physicists made a point of

discreetly helping individuals like Meitner: encouraging foreign colleagues to offer invitations to colloquia; making sure that such letters emphasized that all funds would be paid abroad (so that a visa could be withheld on the grounds that money would have to be taken out of Germany); perhaps arranging for such letters to be predated so they appeared to have been sent before any official expulsion from research institutes had taken place. The fact that Hahn did very little of this for his lifelong colleague is not a terrible sin: it just shows that he was not up to the level of the rare, more highly ethical individuals, such as his colleague Strassmann.

What's more serious—or at least, what seems explicable only by Hahn's realizing he'd done something very wrong—was the way Hahn tried to rewrite the history of his relation with Meitner after the war: treating her as having been some sort of junior assistant when he was interviewed by Swedish newspapers in Stockholm in the week before his Nobel Prize ceremony in 1946; later, giving mocking, almost sighing references to how foolishly misguided her attempted advice had been. Meitner suspected it was all a way of Hahn exculpating himself—for if she'd hardly been there at all, how could he be charged with having treated her badly? See Sime, *Lise Meitner*, Chapters 8 and 14, and especially her note 26 on page 454.

104   ". . . in the lurch": Ibid., p. 185.

104   "Hahn says I should not . . .": Ibid.

105   Hahn, as ever, seemed the slowest . . . : After the outlines of fission had been worked out, he still had troubles: "Bohr will perhaps think I'm a cretin," Hahn wrote to Meitner in July 1939, "but even after 2 of his long explanations I again don't understand it." As with Lawrence, though, the question is one of degree. Hahn was an intelligent enough man—he just wasn't at Meitner's level. What he was exceptionally good at, though, was judging when a field was ripe. That's indispensable. It was not entirely by chance that he "happened" to be at Rutherford's Montreal lab, just when it was possible for a skilled chemist like himself to discover a new element; nor that

he was at the new institutes on the edge of Berlin when those were the most fruitful places for a chemist with his background to be.

Peter Medawar called this importance of appropriate selection "the art of the soluble". The point is not that only easy problems are targeted; rather, "the art of research [is] the art of making difficult problems soluble by devising means of getting at them." Einstein when young was superb at this; Rutherford kept the ability his whole life. The Medawar quote is on page 2 of his justly lauded *Pluto's Republic* (Oxford: Oxford University Press, 1984).

105   "Meitner's opinion and judgment . . .": Watkins, p. 185.

106   Hahn's letter of December 19, 1938, to Meitner: Sime, *Lise Meitner,* pp. 233-34.

106   "Meitner was the intellectual leader . . .": Ibid., p. 241.

106   "You see, you will do a good deed . . .": Ibid., p. 234.

107   Robert Frisch: Many texts speak of someone named Otto Frisch, who confusingly seems to have been related—nephew perhaps?—to an earlier physicist named Robert Frisch. They're one and the same. As a young man, Robert Otto Frisch had used his first name, but when he later ended up working with Americans, for whom the name Robert was so common, Frisch decided being known by his middle name would be less confusing.

107   "Fast, but not auntie": Otto Frisch, *What Little I Remember* (Cambridge: Cambridge University Press, 1979), p. 33.

107   The next morning, when he came down . . . : What happened at breakfast and then during their famous walk in the snow has been extensively recounted by the two participants. See the Frisch and Meitner items in the Guide to Further Reading, as well as the bibliographic notes at Sime, *Lise Meitner,* p. 455, and Richard Rhodes, *The Making of the Atomic Bomb,* p. 810, entry 257.

108   ". . . that she could walk just as fast without." Frisch, *What Little I Remember,* p. 116.

108   "so overwhelmingly new and surprising": Lise Meitner, "Looking Back," *Bulletin of the Atomic Scientists* (Nov. 1964), p. 4.

110   "Fortunately [my aunt] remembered . . .": Frisch, *What Little I Remember,* p. 116. Meitner knew this from earlier published measurements of nuclear weights.

112   The label *fission* . . . : The biological analogy was a common one: Rutherford had chosen the word *nucleus* for the center of an atom on the same basis.

10. Germany's Turn

117–118   Einstein's letter of August 2, 1939: The letter is in most Einstein biographies or histories; see, e.g., *Einstein: A Centenary Volume,* ed. A. P. French (London: Heinemann, 1979), p. 191, for a clear facsimile. The story of how Einstein came to sign the letter is given with exuberant detail in Leo Szilard, *The Collected Works* (Cambridge, Mass.: MIT Press, 1972), though with somewhat more accuracy in Eugene Wigner, *The Recollections of Eugene P. Wigner* (as told to Andrew Szanton) (New York: Plenum Press, 1992).

118   Roosevelt's letter of October 19, 1939, in *Einstein on Peace,* ed. Otto Nathan and Heinz Norden (New York: Simon & Schuster, 1960), p. 297.

120   "I received a report . . .": This diary entry jumps ahead from the main narrative; Goebbels made it in 1942, after the February meeting where Heisenberg made a powerful presentation to a number of Nazi officials, explaining how easily one could proceed with a bomb.

121   . . . he'd always faithfully rejected the job offers . . . : See, David Cassidy, *Uncertainty: The Life and Science of Werner Heisenberg* (New York: Freeman, 1992), pp. 412-14.

121   . . . his wife later said he had nightmares . . . : Ibid., p. 390.

122   "Oh, you know, Mrs. Himmler . . .": Alan Beyerchen, *Scientists Under Hitler* (New Haven, Conn.: Yale University Press, 1977), pp. 159-60. The interview with Beyerchen took

place 34 years after the events; possibly Heisenberg's playing up his mother's naïveté.

122 "Very Esteemed Herr Professor . . .": The letter is reproduced in Samuel Goudsmit, *Alsos: The Failure in German Science* (London: Sigma Books, 1947), p. 119.

125 What would have been a near miss . . . : This is the operation of the famous Uncertainty Principle, which had been worked out largely by Heisenberg in the mid-1920s. It's an odd effect—but central to how $E = mc^2$ came to finally be removed from the laboratory, and turned into such an overpowering force on Earth. It's also, like $E = mc^2$, one of those immensely powerful equations that can be written out in a brief space; in its essence it's simply $\Delta x.\Delta v \geq h$. The Δx is the inexactitude in measuring where a particle is, and Δv is the inexactitude in measuring the velocity at which it's moving. (The symbol "h" is the extremely small figure known as Planck's constant.)

What the $\geq$ in the equation says is that reality's accuracy has a little seesaw or teeter-totter built into it. If you start measuring a particle's location more accurately, then you'll start measuring its velocity less accurately, and vice versa. When one goes up, the other goes down.

This has no direct effect on the large objects around us in ordinary life, but on the micro level, and in what Heisenberg was trying to do in 1940, it's crucial. If you slow down a neutron that you're propelling at a target, then you'll be able to measure its velocity more accurately than you could before. By the "teeter-totter" of the Uncertainty Principle, however, this means you won't be able to measure its location as accurately. In symbols, as the Δv gets less, the Δx gets bigger.

That might seem just a matter of clever words, but—as with the curios of relativity in our earlier chapters—it really does come true. Because Δx is larger, there's a greater spread in our possibility of specifying the neutron's location. *That means its interaction with the target changes.* For what is a fruitful definition of an incoming object's size? Simply how likely it is that it'll contact the nucleus it's being shot at.

It can be irritating to think that this is as good a definition of "size" as one can get, but again, think of the way that in special relativity there was no objective background or "true" time within which events could be placed. To feel that there is a "true" size that can be measured is, indeed, in itself a violation of the Uncertainty Principle. Thus a baseball glove or cricket glove allows you to catch balls which otherwise you would have missed: the size of your hand has effectively increased, due to the extra webbing. But if a viewer knew very little about the game, and just caught a quick blur of the catch on a TV screen, it would be just as plausible for the viewer to conclude that it was the ball that had enlarged, and that this was why fielders could suddenly make such spectacular catches.

With the Uncertainty Principle, there's no way of getting past that blur. The incoming neutron has slowed down, the likelihood of there being a "catch" has gone up—and that's as much explanation as we're going to get of why the target has "become" larger. (In real life the Principle is probabilistic, and the effective "widening" only applies to a sequence of neutrons being shot out.)

The Uncertainty Principle was fundamental to how $E=mc^2$ was released, for it was used in many other calculations needed to construct a bomb. (The electrons in an atom, for example, can't be going too fast—they'd fly away, otherwise—but that constraint on their speed means there's a decreased detail available for any calculations about their actual location within an atom.)

129   "Germany has actually stopped the sale . . .": The letter is widely quoted. See, e.g., *Einstein: A Centenary Volume*, ed. A. P. French, p. 191.

129   But Heisenberg had a procurement organization . . . : Much of the timing is clarified in Mark Walker's *German National Socialism and the quest for nuclear power 1939–1949* (Cambridge: Cambridge University Press, 1989); see especially pp. 132–3. It was in 1943 that the women were "bought" from Sachsenhausen; at the same time, Russian prisoners of war were being used in other aspects of the bomb project (they were forced to work, for example, on Bagge's isotope sluice). Late in the war, when parts of the

Kaiser Wilhelm Institute for Physics was being relocated to the Hechingen area, Heisenberg was informed that Polish slave labor would be available.

129 . . . female "slaves": With the passage of time, it is easy to forget what attitudes the individuals who worked in Germany during the war were accepting; what words like *bought* and *slave* actually meant. There are tens of thousands of pages in the Nuremberg Trial documentation; on November 15, 1947, the New York *Herald Tribune* reported the firsthand testimony behind just one:

> *Nuremberg, November 14, 1947 (A.P.) A French witness testified today that the I. G. Farben combine purchased 150 women from the Oswiecim [Auschwitz] concentration camp, after complaining about a price of 200 marks (then $80.00) each, and killed all of them in experiments with a soporific drug.*
>
> *The witness was Gregoire M. Afrine. He told the American military tribunal trying 23 Farben directors on war crimes charges that he was employed as an interpreter by the Russians after they overran the Oswiecim camp in January 1945 and found a number of letters there. Among the letters, he said, were some addressed from Farben's "Bayer" plant to the Nazi commandant of the camp. These excerpts were offered in evidence:*
>
> *1. In contemplation of experiments with a new soporific drug, we would appreciate your procuring for us a number of women.*
>
> *2. We received your answer but consider the price of 200 marks a woman excessive. We propose to pay not more than 170 marks a head. If agreeable, we will take possession of the women. We need approximately 150.*
>
> *3. We acknowledge your accord. Prepare for us 150 women in the best possible condition, and as soon as you advise us you are ready we will take charge of them.*
>
> *4. Received the order of 150 women. Despite their emaciated condition, they were found satisfactory. We shall keep you posted on developments concerning the experiment.*
>
> *5. The tests were made. All subjects died. We shall contact you shortly on the subject of a new load.*

129   Heisenberg had expressed his impatience . . . : On Heisenberg's sense of urgency, see, e.g., Cassidy, *Uncertainty,* pp. 428–89.

130   "I have now learned that research . . .": *Einstein on Peace,* Otto Nathan and Hans Norden (New York: Schocken, 1968), p. 299.

130–31   ". . . this office would not recommend the employment of Dr. Einstein . . .": Richard A. Schwartz, "Einstein and the War Department," *Isis,* 80, 302 (June 1989), pp. 282–83.

131   Lawrence was not especially bright . . . : Again, as with Hahn, brightness is a relative matter. Lawrence understood his own limitations—"You've got to practically crucify yourself if you're going to amount to anything" he'd explained to an assistant when he was first teaching at Berkeley (in Nuel Phar Davis, *Lawrence and Oppenheimer,* London: Jonathan Cape, 1969, p. 16)—and, in part, as a result Lawrence was exceptionally keen on tracking outside developments which he could incorporate for his own work. His great success was improving a Norwegian's method for accelerating charged particles—it was the basis of the cyclotron, and ultimately what won him his Nobel Prize. Such anxious "borrowing" is central to one sort of successful lab: See Kealey's *Economic Laws of Scientific Research* in the Guide to Further Reading for Chapters 8 and 9.

131   "a tiny cube of uranium . . .": Davis, *Lawrence and Oppenheimer,* p. 99.

132   . . . a practical engineering degree . . . : Wigner, *Recollections of Eugene P. Wigner* (New York: Plenum Press, 1992), pp. 59–62. The caution was widespread: even the intellectually awesome von Neumann took a chemical engineering qualification along with his doctorate in mathematics summa cum laude. Einstein also kept involved in practical inventions—a better electrical current monitor; an improved refrigerator—in part for similar reasons.

132   What shape . . . should the uranium be . . . : The points are presented on p. 40 of Jeremy Bernstein's intro-

duction to his *Hitler's Uranium Club*. Bernstein's work has been central to my understanding of German work on the bomb; I've drawn on it throughout these chapters. Note that when other teams in Germany showed that cubes were in fact the more efficient shape, Heisenberg resisted their findings till most of the war was over—much like his suppression of the views of German researchers who argued for moderators other than the heavy water he preferred.

132   ... flat surfaces ... have the easiest properties to compute . . . : It's a common weakness. The F-117 Stealth fighter, for example, has sharp angular lines not because they're especially aerodynamic—they're not—but because the 1970s computers that were used to analyze its properties couldn't handle anything more rounded. See Ben Rich and Leo Janos, *Skunk Works: A Personal Memoir of My Years at Lockheed* (Boston: Little, Brown, 1994), p. 21.

133   The United States had an army . . . below the tenth rank . . . : How much a backwater America still was—both intellectually, and militarily—in the late 1930s is often overlooked. If anything, it was the U.S. experience in running such administrative/military efforts as the Manhattan Project that contributed to the triumphantly confident postwar view.

II. Norway

134   ... already was a perfectly sound heavy water factory ... [in] Norway: It was actually a fertilizer factory, attached to a large hydroelectric installation. When hydrogen and oxygen are separated to make fertilizer, it's easy to accumulate heavy hydrogen. The heavy water was built up from that.

135   It was a fateful decision . . . : Academic families in pre-1945 Germany were often among the most nationalistic, identifying easily with the militarily-proud Berlin government. Many of these families saw Germany's rise as dependent on such "heroic" moves as the attacks on Denmark and Austria in the 1860s and on France in 1870, and the invasion of Belgium in 1914.

When those expansions collapsed in 1918 the feeling of being trapped simply got stronger. There were constant reminders: When Heisenberg was dominating the world's physics establishment with his 1920s work in quantum mechanics, French occupation troops—often of the lowest quality—were still on his nation's soil. The result was a querulous, resentful tone in much of the country's elite—and so a burst of satisfaction, when finally, in the first successful years from 1936 on, the long-delayed expansion could begin again.

135 "democracy can't develop sufficient energy . . .": Cassidy, *Uncertainty*, p. 473; indeed all of Chapter 24 are recommended. See also, e.g., Abraham Pais, *Neils Bohr's Times* (Oxford: Oxford University Press, 1991), p. 483 as well as Walker, *German National Socialism*, pp. 113–115.

138 "It fell to me . . .": R. V. Jones, "Thicker than Water," in *Chemistry and Industry*, August 26, 1967, p. 1422.

139 "Halfway down we sighted our objective . . .": Knut Haukelid, *Skis Against the Atom* (London: William Kimber, 1954; revised 1973, London: Fontana), p. 68.

140 "Where trees grow, a man can make his way": Ibid., p. 65.

12. America's Turn

143 [Ernest Lawrence's] personnel skills made Heisenberg look considerate . . . : Which does not mean that Lawrence's managerial style couldn't produce benefits in another fashion. For Lawrence ended up collecting students who prospered in his sort of environment. Many of them stole from each other, or snipped out each other's names from original copies of experimental results, yet the Berkeley lab was never strictly immoral. It was "amoral"—and that's very different. Many of its members simply raced in any way possible to fulfill what the outside world wanted. If prestige in medicine came from finding tools to cure disease, then that's what they would back-stab to achieve.

When E = mc² and associated technologies opened up a range of new opportunities, Lawrence's squabbling young men became some of the main controllers of the "spigot" carrying those new powers into our world. They supplied the improved medical tracers to de Hevesy and his colleagues; they worked on improved devices for practical X-ray focusing, for radiotherapy in cancer, and much else. After the war the atomic bomb project opened a gushing well—of grants, contacts; technical knowledge—and Lawrence's men simply happened to be very experienced at pressing to the front of whatever well they saw. An entire book could be written on the interplay between the ethical and practical issues involved.

143 America's own physics establishment had been so weak . . . : But this was changing fast. For the way returning postdoctoral fellows seeded prime universities in the U.S., see Daniel J. Kevles, *The Physicists: The History of a Scientific Community in Modern America* (Cambridge, Mass.: Harvard University Press, 1995); especially Chapter 14.

144 "In July 1939, Lawrence . . .": Emilio Segrè, *A Mind Always in Motion* (Berkeley: University of California Press, 1994), pp. 147–48.

144 . . . factories thousands of feet long . . . able to filter toxic uranium clouds: And this—not the space program—is where Teflon got its first commercial use. The pumps controlling the Tennessee factory filters needed scalants that would be immune to the highly reactive vapor. Substances where fluorine atoms wrap protectively around carbon chains are ideal; the resultant polytetrafluoroethylene is what later became shortened to *Teflon*. Eventually it was realized that a substance that toxic uranium clouds weren't going to stick to would have little problem with the frying pan residues of ordinary suburban kitchens. When the same polytetrafluoroethylene is stretched to form a membrane, Goretex is the result.

145 "He said he appeared not to believe in the eventual success . . .": Peter Goodchild, *J. Robert Oppenheimer: Shatterer of Worlds* (New York: Fromm, 1985), p. 80.

146   "It won't be any trouble . . .": Alice Kimball Smith's 1976 interview with Nedelsky, in *Robert Oppenheimer: Letters and Recollections,* ed. A. K. Smith and Charles Weiner, (Palo Alto: Stanford University Press, 1995), p. 149.

148   One team . . . was simply trying to pull out the most explosive component of natural uranium: This is the famous U235, which forms a bit under 1 percent of ordinary uranium, whose main ores are the calmer U238. One way to remember the difference is that you can hold 50 pounds of U238 in your cupped hands, and only feel a slight warmth, but if you ever found two separate 25-pound chunks of U235 and decided to bring them together, the best details your next of kin could hope to get would have to be supplied by CNN helicopter-borne camera crews, using extreme telephoto lenses to get pictures of the blast site and crater.

A more humdrum way to remember the difference between these two types of uranium is by focusing on the nature of even and odd numbers. Since U238 has 238 particles in its nucleus, everything inside that nucleus is "paired off": an incoming neutron isn't going to have any loose partner to affect easily. But since U235 has an odd number of 235 particles in its nucleus, that means there are 46 pairs of protons and 71 pairs of neutrons—and one extra neutron. That's the vulnerable one. When a fresh neutron arrives from the outer world, it easily reacts with the spare neutron; the result is now 46 tightly bound pairs of protons and 72 tightly bound pairs of neutrons. When a nucleus is configured in this "tighter" way, it's much easier for potentially fissile segments to shoot out. Why that happens—and how it produces a lower energy barrier—is at the heart of practical atomic engineering.

148   Although . . . there were exceptions . . . : The Du Pont engineers who constructed the setting for the Hanford reactor core knew little of atomic physics, but they did know the basic engineering principle that something's always going to go wrong, and you need to allow extra architectural space for the fixes. When the first full running of the reactor slowed due to xenon building up as a by-product of the reaction, they had left enough extra space—following

Wheeler's earlier suggestion—that it was easy to increase the amount of uranium used without tearing apart and rebuilding the reactors. The extra uranium's power more than made up for the xenon. See John Archibald Wheeler, *Geons, Black Holes, and Quantum Foam* (New York: Norton, 1998), pp. 55–59.

150   . . . a ball of plutonium . . . low-density: The phrase *low-density* is of course relative; it's still far denser than lead. The significant point is that it's not dense enough to self-ignite.

150   "It stinks!": Nuel Phar Davis, *Lawrence and Oppenheimer* (London: Jonathan Cape, 1969), p. 216.

150   Teller was vain enough . . . : Teller's private project was the hydrogen bomb, a device far more powerful than what could be built out of uranium. The fact that Oppenheimer later had doubts about its necessity was one of the reasons a petulant Teller testified against Oppenheimer in post-war loyalty hearings.

151   "All that day Serber amused herself . . . ": Serber, *The Los Alamos Primer* (Berkeley: University of California Press, 1992), p. 32. From the same page: "I remember someone at Los Alamos saying that he could order a bucket of diamonds and it would go through Purchasing without a question, whereas if he ordered a typewriter he would need . . . to get a priority number and submit a certificate of need."

151   "It is possible . . .": Richard Rhodes, *The Making of the Atomic Bomb* (New York: Simon & Schuster, 1986), pp. 511–12. I've added the layout of addressee and date.

152   Even a few pounds . . . uninhabitable for years: What could Germany have plausibly achieved? Probably not an entire bomb, but a reactor using carbon dioxide as a moderator rather than heavy water had been strongly pushed by Paul Harteck, the physical chemist based in Hamburg. It would have been easy to construct with the uranium supplies and engineering skills Germany had; the large amounts of highly radioactive substance produced would have been simple to mount on a V-1 or V-2. Note that Otto

Skorzeny seems to have proposed launching a radioactive weapon from a submarine to explode in New York. Coming from ordinary staff planners, that proposal could have been discounted, but Skorzeny was the man who'd organized and led the glider-borne assault that snatched Mussolini from an "impregnable" mountain prison in 1943. Certainly Nazi submarines could easily reach the East Coast of the United States, and occasional ones had been equipped to launch small planes.

Most of all, though, a reason for ongoing caution was the extraordinarily deep engineering and scientific establishment that Germany still had, even in the midst of the war. America avidly employed any chemists with experience in the Clusius process used for separating isotopes, but Germany had Professor Clusius himself—as it also had Professor Heisenberg, Professor Geiger, and the rest. There was a huge middle ground of production engineers, able to pull out such surprises as the factories of jet-powered and rocket-powered aircraft, the extreme long-range submarines, the V-2 rockets, and other devices available before the end of the war. Many of those had problems being produced and deployed in large numbers, but a reactor or even a complete bomb that Heisenberg had managed to finish would only have needed to be deployed once or twice to possibly change the decisions of entire nations.

How close could it have come? In early 1940, Harteck felt he'd need up to 300 kilograms of uranium to test his carbon dioxide idea. He arranged to get the dense frozen carbon dioxide (dry ice) from I. G. Farben; a train car (from army ordnance) to speed it to Hamburg; the necessary uranium from Heisenberg and the Auer company. But at the last moment, Farben declared they could only supply the dry ice until early June; they'd need it after that for keeping food fresh during the hot summer months.

Harteck was frantic, but he could only get the full uranium amounts from Heisenberg in late June. Farben wouldn't budge. Harteck scraped together about 200 kilos of uranium, but with that low amount his results were inconclusive; Germany did not go ahead with the easy, dry ice reactor that (later experience shows) would almost certainly have given them plenty of radioactive metal early on

in the war. Thus was the clear hot weather of that summer—so often cursed by the Allies for letting Panzer armies
advance into France—central to forestalling this greater
evil. Mark Walker, *German National Socialism and the quest for
nuclear power 1939–1949* (Cambridge: Cambridge University
Press, 1989), p. 25 passim has Harteck's efforts; see also
Bernstein, *Hitler's Uranium Club*.

152 . . . Eisenhower accepted Geiger counters . . . : Groves
met with General Marshall on May 23, 1944 and explained
that "Radioactive materials . . . are known to the Germans;
can be produced by them and could be employed as a military weapon. These materials could be used without prior
warning in combating an Allied invasion of the Western
European Coast." The meeting led to portable Geiger
counters being produced and a mission to Eisenhower explaining their use. Directives soon went out from Eisenhower's England-based staff that any invasion force
officers finding strangely fogged X-ray film were to immediately report this to GHQ, and similarly any units that experienced a strange new epidemic disease "of unknown
etiology" having symptoms of hair loss and nausea. See
Leslie Groves, *Now It Can Be Told: The Story of the Manhattan
Project* (London: Andre Deutsch, 1963), pp. 200–203.

153 George de Hevesy had dissolved [the Nobel gold
medals] . . . : George de Hevesy, *Adventures in Radioisotope Research* (London: Pergamon, 1962), p. 27

154 . . . no German divers could bring it up from the
lake's depths: The American Great Lakes are fairly shallow
dips in the ground, where glaciers scraped away the surface, but Tinnsjö is a sheer mountain valley over 1,000 feet
deep that has filled with water. It's one of the deepest lakes
in Europe.

154 Norway command to London: The radio messages
were recalled from memory, in Haukelid's *Skis Against the
Atom*, p. 126. I've changed the heading from "Hardanger
command" to "Norway command," and added "stop" between sentences (as Haukelid did for an earlier message on
p. 78). The Hardanger plateau was the region where the
men were operating.

155 "When I left the watchman . . .": Haukelid, *Skis Against the Atom*, p. 132.

159 The equipment lugged from Berlin . . .": For the cave's siting: Boris Pash, *The Alsos Mission* (New York: Award Books, 1969), p. 206ff; also David Cassidy, *Uncertainty: The Life and Science of Werner Heisenberg* (Freeman, 1992), p. 494. For waiting till dawn on Heligoland: Werner Heisenberg, *Physics and Beyond: Encounters and Conversations* (London: George Allen & Unwin, 1971), p. 61.

159 . . . the German researchers had reached about half the rate . . . : They'd achieved a neutron multiplication rate of nearly 700 percent (in Heisenberg's recollection). About twice as much—requiring more uranium and more heavy water—would have been needed for a sustained reaction. See Cassidy, *Uncertainty*, p. 610.

160 By August, [U.S. bombers] had burned out fifty-eight [Japanese cities]: The gasoline and napalm that American fliers poured onto Japanese cities didn't have enough energy to achieve such destruction. The real source was the thermonuclear radiation of the sun. That radiation beamed down to Earth, and accumulated over years in the chemical bonds within trees. The Japanese themselves had collected that energy—in the lumber with which their cities were largely built. What the American firebombs did was briefly lower the barriers keeping that originally thermonuclear—and now chemical—energy inside. In other words, once sufficient ignition had been created, the firestorms powered themselves.

161 "I told him I was against it on two counts . . .": Harold Evans, *The American Century* (London: Jonathan Cape, 1998), p. 325.

162 Notes from the Presidential Interim Committee . . . : Richard Rhodes, *Making of the Atomic Bomb* (New York: Simon & Schuster, 1986), pp. 650–51. Further detailed minutes of the meeting are available in Martin J. Sherwin's *A World Destroyed: The Atomic Bomb and the Grand Alliance* (New York: Knopf, 1975), see especially pp. 302–3.

162   The army built weapons to use them: Scientists in many countries have similar memories about being abruptly reminded where they came in the chain of command. Andrei Sakharov—a great physicist before he became a dissident—described the evening after the Soviet Union's powerful thermonuclear test of 1955. Marshal Nedelin gave a banquet that night for all the top staff involved. And Sakharov, recalling the spreading fireball he'd seen, gave a toast:

> "I said something like: 'May all our devices explode as successfully as today's, but always over test sites and never over cities.' The table fell silent, as if I had said something indecent. Nedelin grinned a bit crookedly. Then he rose, glass in hand, and said: 'Let me tell a parable. An old man wearing only a shirt was praying before an icon. "Guide me, harden me. Guide me, harden me." His wife, who was lying on the stove said: "Just pray to be hard, old man, I can guide it in myself." Let's drink to getting hard.'
>   'My whole body tensed, and I think I turned pale . . . I drank my brandy in silence . . . Many years have passed, but I still feel as if I had been lashed by a whip . . ." Andrei Sakharov, in *Memoirs*, trans. Richard Laurie (London: Hutchinson, 1990), p. 194.

13. 8:16 A.M.—Over Japan

163   "an elongated trash can with fins": The main sources for the account in this chapter are Rhodes, *The Making of the Atomic Bomb*, pp. 701–15; Robert Serber, *The Los Alamos Primer* (Berkeley: University of California Press, 1992), especially pp. 35–49; as well as standard textbook physics. The "trash can" observation was made by Jacob Beser, one of the plane's crew members; in Rhodes, p. 701.

164   At just under 2,000 feet the height would be ideal: There is a great hardening of attitudes in wartime. In the memo which Frisch and Peierls wrote in March 1940, detailing the theoretical possibility of a practical atomic bomb, they observed:

Owing to the spread of radioactive substances with the wind, the bomb could probably not be used without killing large numbers of civilians, and this may make it unsuitable as a weapon for use by this country. (Use as a depth charge near a naval base suggests itself, but even there it is likely that it would cause great loss of civilian life by flooding and by the radioactive radiations.)

The point is not that dropping the bomb was necessarily wrong: rather, five years after this civilized memo, altitudes for optimal blast yield over a mostly civilian center had become part of ordinary work. For a full copy of what Frisch and Peierls wrote—and what Briggs locked in the safe—see Rudolf Peierls, *Atomic Histories* (New York: Springer-Verlag, 1997), pp. 187–94. On the way that democracies, especially, are liable to experience such chilling transformations, there are Tocqueville's famous 1830s comments on the careerist factors involved (*Democracy in America*, vol. 1, part 3, chapter 24), while Victor Davis Hanson goes much deeper in his first-rate *The Soul of Battle* (New York: The Free Press, 1999).

165   . . . the ordinary electricity of the protons was available . . . : The power is so great that it's often thought nuclear explosions are produced by some new form of energy, which had never existed before. But that's not so. Atomic bombs simply explode because of static electricity.

Electrical repulsion depends very much on the distance between objects. Hold your finger far from a metal surface on a dry winter day and the force between them won't be enough to break down the resistance of the air between them. But let your finger dangle closer—lowering the distance between them—and the overall force gets bigger, until ZAP!! You get struck by the power of static electricity.

A nucleus is about 1,000 times narrower than a whole atom. This means that each of the charged particles crammed in the nucleus will push apart with about 1,000 times greater force than we're used to from the ordinary pushing apart of more widely spaced surface electrons. (The details are slightly different from this, but the results are similar.)

At the same time, instead of just one charged particle pushing against a single other charged particle—as with two electrons pushing apart—the nucleus of a uranium atom contains 92 charged particles. Normally they're held together by the strong nuclear force, but if that's overcome suddenly, its as if there were 92 charged particles hovering right next to each other, with nothing but crackling electrical repulsion acting between them. Now, when one electron is held near one other electron and starts pushing apart, the energy due to their charge is 1×1. When there are 92 protons, this part of their opposing energy is 92×92, which is over 8,400.

In an atomic bomb, both effects are operating at once. The charged particles in the uranium nucleus push outward about 1,000 times more than in ordinary sparks or chemical explosions, from having been crammed into the small nucleus. That gets multiplied a further 8,400 or so times from the charge of the dense, proton-packed nucleus. The total energy forcing itself outward is then on the order of $1,000 \times 8,400$, or more than 8 million times greater than the ordinary electrical forces we're used to, be it the resistance of a wooden bat against a pitched ball or the wild roar of a rocket's chemical fuel exploding. A full calculation requires more adjustments, but the overall proportions are accurate enough. It might sound like hyperbole to say that an atomic bomb is millions of times more powerful than any previous explosive—but it's true.

165 Mass was . . . coming out as the energy of speeding nuclei fragments: (This and a few of the next entries show how $E=mc^2$ enters into practical atomic engineering and astrophysics.) Most of the exploding uranium over Hiroshima remained as a dispersed haze, and only about one percent of each exploding atom got transformed. It doesn't seem much, for if one takes the mass of a single uranium atom, multiplies it by c^2 ($E=m \times c^2$), and then divides by 100 (to take into account this fact that only 1 percent of it "exploded"), the figure that results is a mere 2.7×10^{-6} ergs. That's far too little energy to even blow out a candle, but there were well over 100,000,000,000,000,000,000,000,000 uranium atoms labori-

287

ously concentrated by American technicians in the Hiro-shima bomb. Those combined microbursts were what killed so many people, shattering so many buildings and roads.

166 [the uranium fragments] soon were . . . traveling at a substantial fraction of the speed of light: In the entry on Newton for Chapter 7, we saw how the power of equations allows a researcher to know the strength of our planet's gravity at the distant orbit of the moon, without ever having to leave a book-lined study on Earth. In the same way, it really is possible to look inside an exploding atomic bomb, and accurately compute the speed of the roaring shattered fragments. The equation that allows this, in part, is none other than the old kinetic energy formula from Leibniz and Emilie du Châtelet.

By their work, the kinetic energy of the speeding fragments is $= 1/2mv^2$, where "m" is the mass of the exploding nuclei, and "v" is the speed at which they hurtle apart. If you know that $E=1/2mv^2$, then you can multiply by 2 to get $2E = mv^2$, and then divide by m to get $2E/m = v^2$, and finally take the square root, to end up with the expression $\sqrt{(2E/m)} = v$. Plug in the right values for "E" and "m," and you'll be able to look inside an exploding atomic bomb, and compute the speed of the escaping fragments.

We know the values for the energy "E" of a single exploding uranium atom: it's 2.7×10^{-6} ergs. Insert that in the formula $\sqrt{(2E/m)}$, and the result is that each flung-out fragment of the bomb's core is traveling at the speed of $v=1.2 \times 10^8$ cm/sec. (Again there are modifications, but the overall reasoning holds.) That's over 2 million miles per hour—which is why the solid block of uranium inside the bomb will very quickly become a sphere of hotter-than-boiling gas, racing outward at this extremely high rate.

It's an important result, for the neutrons that are still emerging from the fissioning nuclei will only be able to work if they can catch up with these escaping fragments. That's the reason slow neutrons, of the sort Fermi first analyzed—and which are so useful in gradually cooking up plutonium—are not of any use once a blast finally begins. To keep an explosion going, the bomb has to be constructed so that the shattered fragments release neutrons that travel faster than the escaping cloud of liquefied and then gaseous uranium. Instead of 3 million mph, they have to emerge at 30 million mph or more—and this is what was happening inside the bomb over Hiroshima.

This is also why commercial reactors can't explode like a full-sized atomic bomb: the slow neutrons they operate with wouldn't be able to catch up with an initial explosion; the chain reaction would stop, and the explosion would fizzle out. In that sense, commercial reactors are intrinsically safe. (Again, though, "safe" is a relative term. Even an incomplete explosion can still tear a generator apart with a mighty bang: the roof of the Chernobyl containment vessel weighed many tons, but was flicked aside like balsa wood when the fuel underneath it overheated.)

The kinetic energy calculation is from Serber's *The Los Alamos Primer*, pp. 10 and 12; points on fast neutrons are succinctly presented in Bernstein's *Hitler's Uranium Club*, pp. 21–22.

166–167 . . . for a brief period . . . conditions similar to those in the early moments of the formation of the universe . . . : Could it have ignited the atmosphere? No, for the heat—though immense—was still not enough to cross the barriers for fusion to start. The only possible candidate for ignition would be the nitrogen that dominates the Earth's atmosphere. But long before the temperatures fusion would require could be reached, electrons would radiate away their energy so quickly that the necessary localized heat could never build up. The popular story that such an ignition might occur seems to have stemmed from a misunderstanding in a 1958 interview of one of the chief administrators by the novelist Pearl Buck. An excellent nontechnical summary of the physics is in Hans Bethe's *The Road From Los Alamos* (New York: Simon & Schuster, 1991), pp. 30–33.

169 . . . $E=mc^2$'s first work on planet Earth was done: There's a famous cover from *Time* magazine, showing a saddened Einstein with a mushroom cloud in the background, and the equation $E=mc^2$ appearing with Biblical authority on the clouds. The causality of Einstein's "re-

sponsibility," however, is more subtle. What happened over Hiroshima accurately followed the equation Einstein had written years before, but the equation was not sufficient for the detailed engineering involved; it was, in a sense, not even "necessary," as nuclear physicists could, in principle, have developed the needed technical expertise without any awareness that the overall pattern being summarized by the equation was taking place.

Despite this, Einstein remained defensive about his link. In a reply to a Japanese newspaper in 1952, he wrote: "My participation in the production of the atomic bomb consisted of one single act: I signed a letter to President Roosevelt." In a 1955 letter to a French historian, Einstein elaborated:

> Now you seem to believe that I, poor fellow that I am, by discovering and publishing the relationship between mass and energy, made an important contribution. . . . You suggest that I should . . . in 1905, have foreseen the possible development of atomic bombs. But this was quite impossible since the accomplishment of a "chain reaction" was dependent on the existence of empirical data that could hardly have been anticipated in 1905. . . . [E]ven if such knowledge had been available, it would have been ridiculous to attempt to conceal the particular conclusion resulting from the Special Theory of Relativity. Once the theory existed, the conclusion also existed.

The popular belief linking his work with the bomb encompasses, I suspect, the awe that even without willing the bomb, Einstein had, in this sense, foreseen it. The quotes are from *Einstein on Peace,* ed. Otto Nathan and Heinz Norden (New York: Simon & Schuster, 1960), pp. 583 and 622–23.

14. The Fires of the Sun

175 "fled down the stairs . . .": All quotes from *Cecilia Payne-Gaposchkin: An Autobiography and Other Recollections,* ed. Katherine Haramundanis (Cambridge: Cambridge University Press, 2nd ed., 1996). "fled down the stairs," pp. 119–20; "bicycling," p. 121; "safely lying on her back," p. 72. "[R]ows of braying young men" is not a direct quote, but the reference is from p. 118.

NOTES

177   . . . the original gas clouds . . . : Not all condensing clouds reach a sufficient density to ignite: the planet Jupiter is one example of such an inrushing cloud that was just a few times too small to achieve thermonuclear burning. It's possible that there are a great number of free-floating planets or larger unignited objects, eternally drifting sunless across our galaxy.

177   "the problem haunted me day and night" and "I expressed to a friend that I liked one of the other girls . . .": *Cecilia Payne-Gaposchkin,* pp. 122 and 111.

178   "I always wanted to learn the calculus . . .": George Greenstein, "The Ladies of Observatory Hill," in *Portraits of Discovery* (New York: Wiley, 1998), p. 25.

179   Her work was more complicated than our example: The new theory was from the Indian theorist Meg Nad Saha. There's excellent background in "Quantum Physics and the Stars. 2: Henry Norris Russell and the Abundance of the Elements in the Atmospheres of the Sun and Stars," by D. V. DeVorkin and R. Kenat, *Journal of the History of Astronomy, 14* (1983), pp. 180–222; for briefer explanations see Greenstein, pp. 15–16, and Payne's autobiography, p. 20. On the extraordinary emergence of individuals such as Saha (and Raman and Bose) in India after 1920—and then their remarkable lack of achievement, after a first burst of world-class work was done—see Chandrasekhar's remarks in Kameshwar Wali, *Chandra: A Biography of S. Chandrasekhar* (Chicago: University of Chicago Press, 1992), pp. 246–53. The breakthroughs, Chandra thought, were part of the prideful self-expression that Gandhi's anti-British resistance encouraged; the subsequent collapse was due to haughty, prickly academic empire building by each suddenly famous researcher—a bane Indian science has suffered ever since.

181   "The enormous abundance [of hydrogen] . . .": *Cecilia Payne-Gaposchkin,* p. 20.

183   [the sun] pumps 4 million *tons* of hydrogen into pure energy each second: How can one possibly work out such things? The hottest noon heat in Death Valley is due to

about one thousand watts of solar radiation hitting a square yard of the Earth's atmosphere directly overhead; if extended to cover the whole planet, that means the total amount of light energy hitting the Earth is 150 quadrillion watts.

To see how much mass is lost within the sun to create that energy for Earth, remember that c^2 is a tremendously large multiplier: We live in such a tiny, "low-speed" niche within the universe that our view of the single mass-energy entity is terribly skewed, so that the "mass" aspect of it seems to loom in the foreground, encompassing tremendous power. Since Energy equals mass times c^2, then mass equals Energy divided by c^2. In other words, $m=E/c^2$. If you substitute 150 quadrillion watts for E and 670 million mph for c, the result is about 4.5 pounds. That's all: The light and heat that arrives on Earth is produced from a mere $4\frac{1}{2}$ pounds of hydrogen going out of existence on the sun.

That, incidentally, is how to work out such figures as the one at the start of this chapter, that the sun explodes the equivalent of so many Hiroshima-sized bombs each second. If the sun were at the center of a huge sphere, with the Earth as just a tiny dot on the inner surface of that sphere, then the full surface area of that sphere would be much greater than that of the Earth. It would be about 2 billion times larger, and since the Sun's fires do spray in all directions, suffusing the entire surface of such an imagined sphere with light, then the amount of mass the Sun "loses" each second is that much greater as well. The amount is eight billion pounds of mass. The bomb over Hiroshima in 1945 achieved its destruction by fully transforming under half a pound of mass into energy, which is how one can conclude that the mass our Sun is exploding into energy each second is equivalent to over 16 billion such bombs.

15. Creating the Earth

185   "The blow was delivered . . .": Fred Hoyle, *Home Is Where the Wind Blows: Chapters from a Cosmologist's Life* (Oxford: Oxford University Press, 1997), p. 48.

185   "I pointed out . . .": Ibid., p. 49.

186   "Each morning, I ate breakfast . . .": Ibid., p. 50.

187 . . . from the faces he saw there . . .: One was Nick
Kemmer, who'd been working on Britain's own atomic
project before he'd suddenly disappeared; another was the
brilliant mathematician Maurice Pryce, who'd also mysteriously vanished from the Admiralty Signal Establishment. See Ibid., pp. 227-28.

188 Implosion was a powerful technique on Earth.: The
overlaps were reflected in recruitment. The head of the theoretical section at Los Alamos, for example, was Hans
Bethe—the same man who, in 1938, had "completed" the
work of Payne and others, perfecting the equations that
describe fusion reactions in the sun.

191 . . . there were hundreds of open-air tests: Which is
how pre-World War I German battleships—or at least parts
of them—have come to land on the moon.

In 1919 the Imperial German battlefleet had surrendered to Britain, and was in the confines of the huge Royal
Navy anchorage at Scapa Flow, up in Scotland. After a
number of months of anxious waiting, the German admiral mistakenly came to believe that the British were about
to seize his fleet. The admiral sent out a priorly agreed-
upon coded signal, and the entire grand fleet scuttled itself. But Scapa Flow isn't especially deep—this is why it was
chosen as an anchorage—and so hundreds of thousands of
tons of high-quality steel was now waiting in those waters,
only a few yards or tens of yards down. In the 1920s and
1930s, portions of the fleet were salvaged: divers welding
the holes, then giant air bladders installed, and some of
the half-submerged giants towed all the way to receiving
docks at Rossyth in the Firth of Forth.

After 1945, what remained took on a special value. It
takes a lot of air to make steel, and all post-Hiroshima steel
has some of the radiation from open-air atomic explosions. Pre-1945 steel doesn't. To this day, three battleships
and four light cruisers from the kaiser's once-grand fleet
rest in Scapa Flow (and intrepid readers can dive to see
them, setting out from Stromness in the Orkneys). There's
no advantage in using them for ordinary purposes—it's
much cheaper to make fresh steel—but for extremely sensitive radiation monitors, as on spacecraft, such pre-

Hiroshima sources are indispensable. Equipment that *Apollo* left on the moon, as well as part of the Galileo probe that reached Jupiter, and even the Pioneer probe now past the orbit of Pluto and on its way to distant star systems, all carry remnants of the kaiser's navy, via this salvaged steel from Scapa Flow. The story is well told by Dan van der Vat, in *The Grand Scuttle: The Sinking of the German Fleet at Scapa Flow in 1919* (London: Hodder and Stoughton, 1982).

192   It's not the most sensible of energy choices . . . : The early cost calculations were also distorted by the belief that since the weight of fuel used would go down by a factor of over 1 million, then generating costs would have to be much lower, at least in some proportion. But fuel is only a small part of an electricity generating station's costs. Firms still need to purchase the land and build the turbines and train the staff and pay their salaries and build cooling systems and install transmission stations and maintain the transmission cables. Many nuclear engineering executives knew they were offering unrealistic cost projections when the first big push for commercial reactors got going in 1960s America; the fact that their designs then had stabilized around a scaled-up version of Rickover's model suitable for the confined spaces of submarines did not add to the merits. In fairness, though, nuclear electricity is free of carbon dioxide emissions (aside from what's involved in ore extraction or site construction), and more recent designs really are fail-safe, making a further Chernobyl event impossible.

16. A Brahmin Lifts His Eyes Unto the Sky

195   In a further 5 billion years, the . . . fuel will be gone: Once again, this is the domain of $E=mc^2$; it allows us to foresee how long our solar system will last. The sun's mass can be symbolized as M^o. Only 10 percent of that is hydrogen in a form available for burning, and as we've seen, only 0.7 percent of *that* will actually transfer "through" $E=mc^2$ and pour out as energy. This means the mass actually used will be 0.007 (1/10) × (M^o), which comes out to 1.4×10^{30} grams.

The total energy we can hope to get from that mass is $E=mc^2$, which in this case is $E=(1.4 \times 10^{30}$ grams$) \times (670$ million mph$)^2$. Multiply it out, and the maximum energy the sun can supply till its fuel is used up—under the assumptions above—is, in common units, 1.3×10^{51} ergs.

How long will that total last? It simply depends on the rate at which it's being used. The sun pours out energy—or "shines"—at the rate of 4×10^{35} ergs each second. (This is the sort of figure computable by the reasoning in the note pegged to p. 135, which worked backwards from the amount of sunlight arriving per square yard.) Multiply the total energy the sun can produce till it depletes itself, by this rate at which the depletion is taking place, and the result is 3.2×10^{17} seconds. When that number of seconds is gone, our sun's existence is over (given the approximations of mass availability and constant luminosity we're using). The Earth will either be burned, or absorbed, or flung loose. In slightly more wieldy units, 3.2×10^{17} seconds is about 10 billion years. Since we're about halfway along in the solar process, that's the reason we can assert there are about 5 billion years left.

196   "Some say the world will end in fire . . . : From *The Poetry of Robert Frost*, ed. Edward Connery Lathem (New York: Holt, Rinehart and Winston, 1969), p. 220.

199   In a small enough star, the buildup of pressure is low enough . . . : In "normal" stars, the extra pressure just forces much of the matter inside to move faster, but in stars already under great pressure, this matter is moving so fast that the energy can't go into raising the speed. As with our imagined space shuttle example from Chapter 5, the energy could only end up increasing its mass. The point is well elaborated in Kip Thorne, *Black Holes and Time Warps: Einstein's Outrageous Legacy* (New York: Norton, 1994), pp. 151 and 156-76; Chandra's reasoning is touched on in Wali, *Chandra*, p. 76.

200   "He was a missionary . . .": Wali, *Chandra*, p. 75.

200   "stellar buffoonery . . .": Ibid., p. 142. Wali's Chapters 5 and 6 give the details of Eddington's attack, as well as its influence on Chandra's later career; see also Chan-

drasekhar's own dignified 1982 remarks, at pp. 130-37 of his *Truth and Beauty: Aesthetics and Motivations in Science* (Chicago: University of Chicago Press, 1987).

202   . . . very little of ordinary matter will be left. . . . : In this book we've mostly looked at $E=mc^2$ as describing a bridge or tunnel that goes in one direction, starting on the mass side and transforming across to energy. But when Robert Recorde drew his typographically innovative '===' in the 1550s, he meant it to be a pathway held open in both directions. Neither side was favored.

   This reverse journey doesn't happen under normal circumstances—shine two flashlight beams at each other and solid objects won't pop into existence and start tumbling from the air. But in the early moments of the universe, temperatures and pressures were so high that pure light did regularly take this reverse journey along the equals sign bridge, and get compressed into mass.

   It didn't occur all at once, as if the universe were a celestial bathtub now suddenly poured full. Much of the newly formed mass kept on exploding back into pure energy. Only when the universe was an aged structure, a ponderous full second or more old, did the transformations stop. But by this time there had been a net accumulation on the mass side of the 1905 equation—and the substance that became the ancestor of us all was in existence. Other considerations were at play as well; the story is well enlarged upon in Alan Guth, *The Inflationary Universe* (London: Jonathan Cape, 1997).

Epilogue. What Else Einstein Did

204   "I was sitting on a chair . . .": *The Quotable Einstein,* ed. Alice Calaprice (Princeton, N.J.: Princeton University Press, 1996), p. 170.

205   "the happiest thought of my life": From an unpublished manuscript Einstein wrote for *Nature* in 1920.

205   "Do not worry . . .": *Albert Einstein, the Human Side,* Helen Dukas and Banesh Hoffmann (Princeton, N.J.: Princeton University Press, 1979), p. 8.

210   "Attention was called . . .": Arthur Eddington, *Space, Time and Gravitation* (Cambridge: Cambridge University Press, 1920), p. 114.

213   "Dear Russell . . .": The telegram from the mathematician J. E. Littlewood appears on p. III of *The Autobiography of Bertrand Russell,* vol. II (London: George Allen and Unwin, 1968).

214   "There was a dramatic quality . . .": The visitor was Russell's collaborator Alfred North Whitehead: *Science in the Modern World* (London, 1926), p. 13.

214   "This is the most important result . . .": Albrecht Fölsing, *Albert Einstein: A Biography* (London: Penguin, 1997), p. 444.

215   Henry Crouch . . . golfing specialist: Meyer Berger, *The Story of The New York Times, 1851-1951* (New York: Simon & Schuster, 1951), pp. 251-52.

217   . . . but English anti-Semites . . . : The quote is from *The Collected Writings of John Maynard Keynes, Vol. X: Essays in Biography* (London: Macmillan; New York: St. Martin's Press, for the Royal Economic Society, 1972), p. 382. The occasion was Keynes's visit to Berlin in June 1926, where he lectured at the University; he met Einstein at a dinner afterwards. "It is not agreeable," Keynes remarked, "to see a civilization so under the ugly thumbs of its impure Jews."

217   ". . . village of puny demi-gods upon stilts": The comment was to his long-time correspondent, Queen Elizabeth of Belgium. See *The Quotable Einstein,* p. 25.

218   "Had I known . . .": Antonina Vallentin, *The Drama of Albert Einstein* (New York: Doubleday, 1954), p. 278.

218   Then, as the years went on . . . : To some extent, what happened to Einstein is a common effect. Great artists and composers often do their top work as they get old, but scientists don't. Partly this could be because it's intellectually too difficult to hold complex ideas in one's head. Even in drama, the play *Oedipus at Colonus,* which Sophocles wrote when very old, has a crudeness of construction that would be not useful in a physical theory. But Beethoven wrote

complex works into his fifties, and *The Tempest* was written in Shakespeare's late forties. There's something more going on in science—and for Einstein, the slippage was more extreme than almost anyone else's. It's a lengthy topic— there are important insights from Macauly and even Spielberg, which we'll explore on the Web site.

218 [A] . . . young assistant once asked him . . . : *Einstein, A Centenary Volume,* ed. A. P. French (London: Heinemann, 1979), p. 32. The assistant was Ernst Strauss, who worked with Einstein from 1944 to 1948. The same volume, p. 211, has Einstein's account of how very different it had been when he was younger and could "scent out the paths that led to the depths, and to disregard everything else, all the many things that clutter up the mind, and divert it from the essential."

218 "Discovery in the grand manner . . .": Banesh Hoffmann, *Albert Einstein, Creator and Rebel* (New York: Viking, 1972), p. 222.

Appendix. Follow-up of Other Key Participants

223 "If I were King . . .": Du Châtelet's preface to her translation of Mandeville's "The Fable of the Bees"; in Esther Ehrman, *Mme du Châtelet* (Berg Publishers, 1986), p. 61.

224 "The gift of leading a harmonious life . . .": *Albert Einstein/Michele Besso, Correspondence 1903-1955,* trans. Pierre Spezialli (Paris: Hermann, 1972), p. 537.

225 "I had then to start . . .": Richard Rhodes, *The Making of the Atomic Bomb* (New York: Simon & Schuster, 1986), p. 356.

228 "A real genius . . .": Rhodes, *The Making of the Atomic Bomb,* p. 448.

228 "The radioactivity was minuscule . . .": Emilio Segrè, *A Mind Always in Motion* (Berkeley: University of California Press, 1994), p. 215.

228-29 . . . deHevesy went back to the jar . . . : *Adventures in Radioisotope Research: The Collected Papers of George Hevesy,* vol. 1 (London: Pergamon Press, 1962), pp. 27, 28.

229   "Why, fellows, you don't want . . .": Nuel Phar Davis, *Lawrence and Oppenheimer* (London: Jonathan Cape, 1969), p. 351.

229–30   "Microphones installed?": Jeremy Bernstein, ed., *Hitler's Uranium Club: The Secret Recordings at Farm Hall* (Woodbury, N.Y.: American Institute of Physics, 1996), p. 75. See also the introduction by Sir Charles Frank to *Operation Epsilon: The Farm Hall Transcripts* (Bristol: Institute of Physics, 1993) for various practicalities of the recording, including his aplomb when questioned about "some unexplained wires in the back of a cupboard."

230   "We have tried to make a machine . . .": Bernstein, *Hitler's Uranium Club,* p. 211.

231   Berlin Auer company: Samuel Goudsmit, *Alsos: The Failure in German Science* (Woodbury, N.Y., 1996), pp. 56-65.

232   "there was literally no time for research": *Cecilia Payne-Gaposchkin: An Autobiography and Other Recollections,* ed. Katherine Haramundanis (Cambridge: Cambridge University Press, 2nd ed. 1947), p. 225.

232   "Icelandic was a minor challenge": George Greenstein, "The Ladies of Observatory Hill," in *Portraits of Discovery* (New York: Wiley, 1998), p. 17.

233   "Fred won't resign . . .": Fred Hoyle, *Home Is Where the Wind Blows: Chapters from a Cosmologist's Life* (Oxford: Oxford University Press, 1997), p. 374.

233   "I am almost ashamed . . .": Kameshwar Wali, *Chandra: A Biography of S. Chandrasekhar* (Chicago: University of Chicago Press, 1992), p. 95.

$$E=mc^2$$

Faraday and Energy

The best way to get to know Michael Faraday as a person is to skim through his collected letters, either the version edited by L. P. Williams et al., *The Selected Correspondence of Michael Faraday* 2 vols. (Cambridge and New York: Cambridge University Press, 1971), or the more comprehensive *The Correspondence of Michael Faraday*, ed. Frank A. J. L. James (London: Institution of Electrical Engineers, ongoing from 1991). There's the teenager racing along the London streets in a rainstorm one night, exulting in the gush of water on his body; later there's the earnest young assistant, furious at Humphry Davy's wife for treating him as a servant on a trip to the Continent; finally, decades after, we see the grand old man of British science, distraught at realizing his memory is fading ever faster, and that the concentration he'd once been able to bring to bear on any subject is now gone.

There's a good look at the way religion entered into his life and approach, in Geoffrey Cantor's *Michael Faraday, Sandemanian and Scientist: A Study of Science and Religion in the Nineteenth Century* (London: Macmillan; New York: St. Martin's Press, 1991) while *Michael Faraday: His Life and Work*, by Silvanus P. Thompson (London: Cassell, 1898), is my favorite overall biography, catching the tone of the times in a way later writers find difficult. The more recent

Faraday Rediscovered: Essays on the Life and Work of Michael Faraday, ed. David Gooding and Frank A. J. L. James (London: Macmillan, 1985; New York: American Institute of Physics, 1989) corrects a number of Thompson's errors, and also is a good introduction to the major scientific discoveries. One of the most exciting chapters annotates an almost minute-by-minute account, based on Faraday's notebooks, of that crucial September 1821 experiment.

Humphry Davy: Science and Power, by David M. Knight (Oxford, England: Blackwell, 1992) is crisply analytic on Faraday's troubled mentor, and on the indispensable role the poets Wordsworth and Coleridge played in shuttling Kant's ideas to Davy and then to Faraday. He also shows the way British scientists were often disposed to believe in mysterious powers—as with Faraday's God-granted unities—to distinguish themselves from the extreme materialists in France, whose research was held to have helped justify the Terror in the French Revolution. For a more flowing account of Faraday and Davy, try *The Mercurial Chemist,* by Anne Treneer (London: Methuen, 1963).

Faraday was far from being the only individual active in developing the conservation of energy. Thomas Kuhn has a famous essay about it, "Energy Conservation as an Example of Simultaneous Discovery," in his *The Essential Tension: Selected Studies in Scientific Tradition and Change* (Chicago: University of Chicago Press, 1977). Kuhn doesn't just vaguely note that conservation was "in the air," but shows the role of the era's many new-fangled industrial machines as a source of metaphors, as well as the importance of the many newly practical technologies for converting between types of energy. *The Science of Energy: A Cultural History of Energy Physics in Victorian Britain,* by Crosbie Smith (London: Athlone Press, 1998) takes a different approach, looking, for example, at the minutiae of Scottish theology and its patronage networks, and how the less stratified social structure there naturally brought engineers, professors, and theologians into a powerfully cross-fertilizing contact.

The extrascientific motivations of the seminal Hans Christian Oersted are explored in R. C. Stauffer, "Speculation and Experiment in the Background of Oersted's Discovery of Electromagnetism," *Isis, 48* (1957), and, if you can bear wading through hundreds of pages of treacle, there are Oersted's own writings. Gerald Holton's essay "The Two Maps" (in his *The Advancement of Science, and Its Burdens* [Cambridge, Mass.: Harvard University Press, 1986, 1998], pp. 197–208) is excellent on the significance of Oersted's being misunderstood.

For a closer look at the science of energy as it has ramified today, the oddly indirect operational definition that has been so useful is made clear in Chapter 3 ("The Great Conservation Principles") of *The Character of Physical Law* (London: Penguin UK, 1992), a transcript of BBC-recorded Cornell lectures by the ever-ebullient Richard Feynman. The notion of entropy is analyzed—with steadily increasing order—in Peter Atkins's estimable *The Second Law: Energy, Chaos, and Form* (New York: Scientific American Books, 1984, 1994), a text excellent at showing the reasoning that led to seeing a further level of structure in energy's operations. (The chapter where he shows how the life we're used to is just a brief stopping point in the full temperature scale of the universe is a masterpiece.) For if we can understand the disorder we call heat, then we should be able to understand the opposite of disorder, which could be called "information." Neil Gershenfeld's *The Physics of Information Technology* (New York: Cambridge University Press, 2000) is at a more advanced level than Atkins, but strongly recommended for anyone interested in exploring this furthermost ramification of the Victorian energy concepts.

Lavoisier and Mass

Lavoisier found a graceful biographer in Arthur Donovan, with his *Antoine Lavoisier: Science, Administration, and Revolution* (Oxford, England: Blackwell, 1993). Jean-Pierre Poirier wrote a more comprehensive book, *Lavoisier: Chemist, Biologist, Economist* (English translation) (College Park, Penn.: University of Pennsylvania, 1996), but it's harder to read straight through. For more than thirty years, Robert Darnton has been probing life under polite society's surface in France around the period when Lavoisier worked, and his

Mesmerism and the End of the Enlightenment in France (Cambridge, Mass.: Harvard University Press, 1968) is excellent for background, especially on the popular attitudes that proved to be fatal for Lavoisier later on. For Marat I'd go back to the brief account *Jean Paul Marat: A Study in Radicalism* (orig. 1927, reissued Chicago: University of Chicago Press, 1967) by the then young Louis Gottschalk. Readers near a large library, and with even a little French, will be gripped by the firsthand accounts of the imprisonment and trial, in Adrien Delahante, *Une famille de finance au XVIIIe siècle*, 2 vols (Paris, 1881).

Stephen Toulmin and June Goodfield's *The Architecture of Matter* (London: Hutchinson, 1962) is especially thoughtful in disentangling some of the attitudes of Lavoisier's time, while the classic *The Origins of Modern Science 1300–1800,* by Herbert Butterfield (orig. London, 1949) take a more no-nonsense frontal approach. A more physics-oriented history, bringing the story into the twentieth century, is Max Jammer's *Concepts of Mass in Classical and Modern Physics* (New York: Dover, 1997), which includes such nutritious tidbits as the plausibilities behind the suggestion that the word *mass* originated in the Hebrew word *matzoh;* also the links between mass conservation, and the vision of *quantitas mataeriae,* or "quantity of matter," which Aquinas's followers had used to resolve the problems of what actually happens during transubstantiation in the Catholic mass.

Of the recent approaches, Frederic Holmes is always fresh, as in "The Boundaries of Lavoisier's Chemical Revolution," in *Revue d'Histoire des Sciences, 48* (1995), pp. 9–48. Maurice Crosland's chapter "Chemistry and the Chemical Revolution" in *The Ferment of Knowledge,* George S. Rousseau, ed. (New York: Cambridge University Press, 1980) is good on what Lavoisier was likely to be thinking during his metal experiments. Try also Perrin's essay "The Chemical Revolution: Shifts in Guiding Assumptions," pp. 53–81 in *The Chemical Revolution: Essays in Reinterpretation* (special issue of *Osiris,* 2nd series, 1988).

The question of what mass "really" is brings us to the concept from modern physics of the Higgs field, for which *Lucifer's Legacy—The Meaning of Asymmetry,* by Frank Close

(New York: Oxford University Press, 2000) is one excellent start, while Gerard 't Hooft's *In Search of the Ultimate Building Blocks* (Cambridge: Cambridge University Press, 1997) gives even wider background, deftly tracking the story through the author's own schooling and professional puzzlements (though modesty—plus the annoying lack of efficient time-travel—forbade his mentioning the story's culmination, for now, in his own Nobel Prize).

"c"

Galileo lived in a time when science hadn't fully separated out from philosophy or literature, which means that today's nonspecialist can have the treat of entering into his major works directly: great chunks of his *Two New Sciences* are immensely attractive to read. I. B. Cohen's old article "Roemer and the First Determination of the Velocity of Light," *Isis,* 31, 1940, pp. 327–79, expands the story of what Cassini had to put up with, while a skillful update, including points on the caution against straying from hard empiricism that Cassini had to follow due to working in a Catholic country in the century of Galileo's persecution, is in "The Galilean satellites of Jupiter from Galileo to Cassini, Römer, and Bradley," by Suzanne Débarbat and Curtis Wilson in *The General History of Astronomy, Vol. 2, Planetary Astronomy from the Renaissance to the Rise of Astrophysics, Part A: Tycho Brahe to Newton,* eds. René Taton and Curtis Wilson (Cambridge: Cambridge University Press, 1989), pp. 144–157. Timothy Ferris's *Coming of Age in the Milky Way* (New York: William Morrow, 1988) puts that within an even wider setting; his book is the ideal introduction to the history of astronomy.

Maxwell found a suitably wry biographer in Martin Goldman, *The Demon in the Aether: The Story of James Clerk Maxwell* (Edinburgh: Paul Harris Publishing; with Adam Hilger, Bristol, 1983). To go with that I'd recommend Chapter 11 ("Gentleman of Energy: the Natural Philosophy of James Clerk Maxwell") in Crosbie Smith's *Science of Energy: A Cultural History of Energy Physics in Victorian Britain* (London: Athlone Press, 1998) as a place to start. As with Ferris, excellent explanations are mixed with pleasing anecdotes;

we even find Maxwell's reflections (heartening to any Oxford author) on his education in Cambridge:

Like a plucked and skinny goose . . I
Asked myself with voice unsteady,
If of all the stuff I read, I
Ever made the slightest use.

Where Maxwell's science comes alive to the general reader is in many of his introductions and commentaries on the major works. Excellent samples are "A Historical Survey of Theories on Action at a Distance," and "Experiment on Lines of Force," conveniently available in *Physical Thought: An Anthology,* ed. Shmuel Sambursky (London: Hutchinson, 1974). The 1890 edition of Maxwell's papers is superseded by *The Scientific Letters and Papers of James Clerk Maxwell,* ed. P. M. Harman (New York: Cambridge University Press, 1990, 1995). For overviews, Harman's classic *Energy, Force, and Matter* (New York: Cambridge University Press, 1982) might be compared with Robert D. Purrington's even more finely honed *Physics in the Nineteenth Century* (New Brunswick, N.J.: Rutgers University Press, 1997).

Innovation in Maxwell's Electromagnetic Theory, by Daniel Siegel (New York: Cambridge University Press, 1991), takes a very detailed, at times polemical, look at the process of Maxwell's creativity, including revealing contrasts with the overtheoretical French tradition; Christine M. Crow's *Paul Valéry and Maxwell's Demon: Natural Order and Human Possibility* (Hull, England: University of Hull Publications, 1972) brings out rich insights from a different examination of the French tradition. Richard Feynman would have had little use for Valéry or most historians, which is a pity, but for exploring the actual science of light, his writing (and research) has had few peers: along with the physics texts in the next entry, *QED: The Strange Theory of Light and Matter,* by Richard Feynman (Princeton, N.J.: Princeton University Press, 1985) is a fine start.

Du Châtelet and "Squared"

Du Châtelet hasn't been favored by her English language biographers, but readers with some French are in for a treat. Elisabeth Badinter had the excellent idea of doing a comparative biography of Emilie du Châtelet and Madame d'Épinay, and her *Émilie, Émilie: l'ambition feminine au XVIIIe siècle* (Paris: Flamarrion, 1983) is a fast-paced, well thought out pairing of psychological portraits.

Les Lettres de la Marquise du Châtelet, 2 vols. (Geneva, 1958), ed. T. Besterman shows du Châtelet at ease, often being funny in the way clever screenwriters are funny—but then she'll shift, almost from one sentence to the next, to a genuine puzzlement about how an observation she's just made would apply to the nature of free will, or affect the foundations of physics.

Voltaire en Son Temps: Avec Mme du Châtelet 1734–1748 (Paris: Albin Michel, 1978) by René Vaillot is more pedantic, but pulls out worthwhile nuggets, such as the tableau of du Châtelet, over morning coffee, impressing a visitor by reading out a letter from Christian Wolff about possible giant inhabitants on the planet Jupiter. The letter was in Latin, and the idea, once developed in conversation with Voltaire, is clearly at the heart of his (highly recommended) short story "Micromégas". Its theme of an innocently wise giant's perspective—whose soul, one suspects, is what Voltaire once hoped for himself—is one that has swirled through texts over the centuries, from the Bible to Hollywood's *The Day the Earth Stood Still* to Ted Hughes's "The Iron Man."

For a straightforward biography *Voltaire in Love,* by Nancy Mitford (London: Hamish Hamilton, 1957) is, as one would expect, not especially accurate in biographical details, clueless about the science, bitchy in tone, and a cracking good read. Fontanelle's *On the Plurality of Inhabited Worlds* (London: Nonesuch Press, 1929), translated by John Glanville, is wonderful on giving a sense of the enthusiasm du Châtelet might have experienced gazing up at night.

The second chapter of *Science and the Enlightenment* by Thomas Hankins (New York: Cambridge University Press, 1985) is the best start for the Leibniz/du Châtelet/Newton issues, while I. O. Wade's *Voltaire and Mme du Châtelet: An Essay on the Intellectual Activity at Cirey* (Princeton, N.J.: Princeton University Press, 1941) is not quite as dry as the title suggests. Steven Shapin's essay "Of Gods and Kings:

Natural Philosophy and Politics in the Leibniz-Clarke Disputes," *Isis, 72,* (1981), pp. 187-215, widens the intellectual struggles, as does, even more, his polemical reinterpretation *The Scientific Revolution* (Chicago, University of Chicago Press, 1996).

Carolyn Iltis's "Madame du Châtelet's metaphysics and mechanics," in *Studies in the History and Philosophy of Science, 8* (1977), pp. 29-48, is a more conventional extending of the historical setting, and fits well with the intriguing "Newtonian Forces and Lockean Powers: Concepts of Matter in Eighteenth-Century Thought," by P. M. Heimann and J. E. McGuire, in *Historical Studies in the Physical Sciences, 3* (1971), pp. 233-306. To situate the château at Cirey as a research institution and show what an intellectually adventurous couple could produce, it would be hard to find more apt authors than Lewis Pyenson and Susan Sheets-Pyenson, joint writers of *Servants of Nature: A History of Scientific Institutions, Enterprises and Sensibilities* (London: HarperCollins, 1999).

Einstein and the Equation

Einstein

I have a weakness for some of the early biographies of Einstein: as with old movies, the very nature of their presentation captures something of the period in which their subject lived, which few currently produced items could match. Two biographies that Einstein himself especially liked are *Einstein: His Life and Times,* by Philipp Frank (New York: Knopf, 1947), his successor at Prague; and *Albert Einstein: A Documentary Biography,* by Carl Seelig, trans. by Mervyn Savill (London: Staples Press, 1956). Seelig was a journalist and friend of the family who corresponded with Einstein for years.

Of more recent works, Banesh Hoffmann's *Albert Einstein, Creator and Rebel* (New York: Viking, 1972) is the ideal introductory mix of biography and scientific background. For the early years, *The Young Einstein: The Advent of Relativity,* by Lewis Pyenson (Boston: Adam Hilger, 1985) shows what thoughtful academic work can achieve, as with Pyenson's hunting out the detailed workings of the family firm Einstein grew up within, and noting his uncle's development of a measuring device that depended on verifying the signals from two independent clocks—a key part, once the issue is pondered over, of the reasoning behind special relativity. Another ingenious probing is in Robert Schulmann's "Einstein at the Patent Office: Exile, Salvation or Tactical Retreat"; in a special edition of *Science in Context*, vol. 6, number 1 (1993), pp. 17-24.

For the cultural setting, very few scientists or historians of science can match the depth of insight which Fritz Stern—one of America's great historians—brings out in the long third chapter of his *Einstein's German World* (Princeton, N.J.: Princeton University Press, 1999), or his earlier "Einstein's Germany" in *Albert Einstein, Historical and Cultural Perspectives*, ed. Gerald Holton and Yehuda Elkana (Princeton: Princeton University Press, 1982), pp. 319-43. One who does reach Stern's heights is Abraham Pais, whose own life has been a mirror of much of what the twentieth century has had to offer, and whose *"Subtle Is the Lord . . .": The Science and the Life of Albert Einstein* (New York: Oxford University Press, 1982) is probably the last account we'll have from a researcher who knew Einstein well. It's built on a close reading of Einstein's papers, so it is more technical than this book, but it's a thorough, reasoned evaluation.

The other standout thinker in Einstein studies is Gerald Holton, who has kept a freshness and depth of insight in his work extending now for more than forty years. I especially recommend his *The Advancement of Science, and its Burdens* (Cambridge, Mass.: Harvard University Press, 1986, 1998), as well as *Einstein, History, and Other Passions* (Reading, Mass.: Addison-Wesley, 1996).

In addition to the Veblen essay, Claude Lévi-Strauss's little pamphlet "Race and History," reprinted in his *Structural Anthropology*, Vol. 2 (New York: Penguin, 1977) elaborates on ways profound ideas can emerge from the colliding of cultures; Mary Douglas's classic *Purity and Danger* (New York: Routledge, 1966) takes a deeper look at the powerful potentialities of conceptual and social fissures. Nilton Bonder's *Yiddishe Kop: Creative Problem Solving in Jewish Learning, Lore and Humor* (Boston: Shambhala Publications, 1999) is an oddly near-mystical account of an intriguing cultural habit, while Howard Gardner's essay "The Creators' Patterns" in *Dimensions of Creativity,* ed. Margaret A. Boden (Cambridge, Mass: A Bradford Book, The MIT Press), pp. 143-58, brings us back to Earth, putting Einstein and Besso in the context of Freud with Fliess, Martha Graham with Louis Horst, and other innovators who needed a supportive friend in their initial years-long period of seeming isolation, even while their later breakthroughs were somehow privately being prepared.

Physics Introductions

For the underlying physics, the ideal thing is to spend a summer with an introductory calculus book, after which all freshman university physics texts suddenly open up. But since life is short, and not everyone has the spare summer, Robert Mills (of Yang-Mills fame) wrote the seemingly easygoing yet quite powerful *Space, Time and Quanta: An Introduction to Contemporary Physics* (New York: W. H. Freeman and Company, 1994), which can provide that freshman-level introduction even for readers who bring no calculus with them.

On a less technical level, an excellent compilation is Timothy Ferris's *The World Treasury of Physics, Astronomy, and Mathematics* (Boston: Little, Brown, 1991). It offers gracefully written essays, often by the key practitioners—there's even a four-page account of $E=mc^2$ by Einstein himself.

The Physics of Star Trek by Lawrence Kraus (New York: Basic Books, 1995) takes another fresh approach: $E=mc^2$, for example, is discussed there in terms of the difficulties a real-life Scottie would face in responding to Captain Kirk's command, "Beam me up." Kraus's later *Fear of Physics: A Guide for the Perplexed* (New York: Basic Books, 1994) develops some of the physics more systematically. *Dance for Two: Selected Essays,* by Alan Lightman (London: Bloomsbury, 1996) is a brightly written account of selected topics: the title essay, for example, describes Newton's laws in terms of the grinding jolts of the whole earth as it moves up or down (ever so slightly!) in response to the leaps of a single ballerina on its surface.

The Strange Case of Mrs. Hudson's Cat: Or Sherlock Holmes Solves the Einstein Mysteries, by Colin Bruce (New York: Vintage, 1998) is the sort of book other authors berate themselves for not having thought of first. Bruce has written a series of Holmes and Watson stories, each of which depends for its resolution on a basic principle from physics. Watson bumbles, Baker Street is fogged in, Professor Challenger is perfidious—and learning is effortless.

Special Relativity Introductions

The Time and Space of Uncle Albert, by Russell Stannard (London: Faber and Faber, 1989), imagines a series of teasing conversations between a kindly Uncle Albert and his trendy niece Gedanken. It's advertised as aimed for teens or even preteens, but it's an excellent start for adults. *Mr. Tompkins in Wonderland,* by George Gamow (various editions), has a similarly sweet touch. Instead of analyzing the whys of the equation, at least at first, Gamow simply places an imaginary befuddled bank clerk within the settings that relativity and other branches of physics describe. (His work has been updated by Russell Stannard in *The New World of Mr. Tompkins* [New York: Cambridge University, 1999]. *Einstein's Legacy: The Unity of Space and Time,* by Julian Schwinger (Basingstoke, England: Freeman, 1986) moves up a level, giving a clear and eloquently written account of relativity and the equation; The Wald and Geroch texts mentioned on page 318 apply as well.

Newton

Of the many Newton biographies, I'd start with A. Rupert Hall's *Isaac Newton: Adventurer in Thought* (New York: Cambridge University Press, 1992). The Norton critical anthology *Newton: Texts, Backgrounds, Commentaries,* ed. I. Bernard Cohen and Richard S. Westfall (New York: Norton, 1995) gives copious extracts from Newton, as well as samplings of the twentieth-century secondary literature, from Keynes and Koyre to Westfall and Schaffer. It's the best guide to going further.

Into the Atom (Chapters 8 and 9)

C. P. Snow's fourteen-page essay on Rutherford in his *Variety of Men* (London: Macmillan, 1968) sounds as if you're hearing an insider whispering to you what really happened at the Cavendish lab in the glory days. There's Rutherford's bluff grandstanding—when told he was always on the crest of the wave, he boomed, "Well, after all, I made the wave, didn't I?!" But there are also the hesitancies underneath, as with Rutherford's quiet, suddenly blurted insistence that certain overseas scholarship funds be continued: "If it had not been for them, I shouldn't have been."

After Snow's essay, try *Rutherford* by A. S. Eve (London: Macmillan, 1939) for further recounting of the early days. Despite the less-than-ingenious title, *Rutherford* by Mark Oliphant (New York: Elsevier, 1972) is an original and intense work, getting across Rutherford's fury—and then embarrassed half-apologies—as he saw the world-dominating research unit he'd created slowly start to break, not least through character flaws of his own. Oliphant was one of the last of Rutherford's promising young students, and the individual who kick-started Briggs to get the U.S. atomic bomb project going; after a distinguished post-war career that included decades of working against nuclear weapons, he died shortly before his ninety-ninth birthday, just weeks before this book was going to press).

The Neutron and the Bomb: A Biography of Sir James Chadwick, by Andrew Brown (New York: Oxford University Press, 1997), is suitably neutral to match the discoverer of the neutron. It goes into the early years thoroughly enough though to show how the quiet Chadwick became one of the only individuals to stand up to both Oppenheimer *and* Groves—thus giving him a key role in the success of the Manhattan Project. The way that the bitter rivalries at the end between Chadwick and his mentor Rutherford were carried out through the mercilessly taut coolness between their wives is best, however, in Oliphant's account.

Atoms in the Family, by Laura Fermi (Chicago: University of Chicago Press, 1954), is an account of Fermi by his wife, who has something of the sweetly teasing tone of Einstein's sister. For more on the scientific background and personality of this quietly driven man, there's *Enrico Fermi, Physicist,* by Emilio Segrè (Chicago: University of Chicago Press, 1970). The evocative essay "Fermi's group and the recapture of Italy's place in physics," in *The Scientific Imagination,* by Gerald Holton (Cambridge, Mass.: Harvard University Press, 1998) goes into the Rome research group in detail, including the importance of Fermi's having found an all-powerful bureaucratic protector.

How did Rutherford and Fermi manage to sustain such powerful research centers? Edward Shils's *Center and Periphery: Essays in Macrosociology* (Chicago: University of Chicago Press, 1975) is good on the standard sociological backing, while J. H. Brown's "Spatial variation in abundance," *Ecology, 76* (1995), pp. 2028–43, is an interesting demonstration of the way *low* competitive pressure can be excellent for fresh speciation. *The Economic Laws of Scientific Research* by Terence Kealey (New York: St. Martin's Press, 1996) takes a quirkily refreshing approach, showing, for example, how pharmaceutical firms and other research groups regularly profit from hiring top scientists who think they're going to do original work, but in fact are useful simply because they can intelligently sieve the available literature.

Lise Meitner: A Life In Physics, by Ruth Lewin Sime (Berkeley: University of California Press, 1996) is clear on the background, and also takes a justly strong feminist line; see also the first-rate essay on Meitner by Sallie Watkins in *A Devotion to Their Science,* ed. Marlene F. and Geoffrey W. Rayner-Canham (Toronto: McGill-Queen's University Press, 1997). They're the ideal backing for Meitner's own brief account "Looking Back," *Bulletin of the Atomic Scientists, 20* (Nov. 1964), pp. 2–7. Otto Frisch's autobiography *What Little I Remember* (New York: Cambridge University Press, 1979) is a delightful view from this gentler man; *Aging and Old Age* by Richard Posner (Chicago: Univeristy of Chicago Press, 1995) is a fresh take on the role of sunk costs in long research careers.

Building the Bomb (Chapters 10-13)

In 1943, armed guards from the United States Army would have taken a strikingly personal interest in any outsider who tried to copy Robert Serber's lectures for arriving scientists at Los Alamos—for those lectures surveyed everything that was then known about building atomic weapons. Copies are now somewhat more conveniently available in Serber's *The Los Alamos Primer* (Berkeley: University of California Press, 1992). Along with all the lectures, now declassified, the book contains Serber's own excellent annotations and reminiscences. It's the ideal way to capture the working mood at Los Alamos.

On Oppenheimer, the best source is *Robert Oppenheimer: Letters and Recollections,* ed. Alice Kimball Smith and Charles Weiner (Palo Alto, Calif.: Stanford University Press pbk 1995; orig. Harvard University Press, 1980). The letters are almost shatteringly direct: there are the brief moments of intellectual joy; then the self-torments, the insecurities, and the layers upon layers of posturing. How he and Lawrence overcame their own warinesses to become best friends—and then exhausted, dismayed enemies—is the drama in Nuel Phar Davis's masterpiece *Lawrence and Oppenheimer* (London: Jonathan Cape, 1969). Richard Feynman's best-selling reminiscences *"Surely You're Joking, Mr. Feynman!",* ed. Edward Hutchings (New York: Norton, 1985) are vivid on the personal level; James Gleick's *Genius: Richard Feynman and Modern Physics* (New York: Pantheon, 1992), gives a far richer story of what Feynman and others experienced on the mesa.

The best overall account of the U.S. and German projects is Richard Rhodes's *The Making of the Atomic Bomb* (New York: Simon & Schuster, 1986), a deserved winner of the National Book Award. Eavesdropping is a guilty pleasure, and in *Hitler's Uranium Club: The Secret Recordings at Farm Hall,* ed. and annotated by Jeremy Bernstein (Woodbury, N.Y.: American Institute of Physics, 1996) we get to eavesdrop on Hahn, Heisenberg, and all the rest of them as the interned German scientists squabble their way through six long months in genteel captivity. Bernstein's background on the science and the personalities is extremely clear. *Alsos: The Failure in German Science,* by Samuel Goudsmit (London: Sigma Books, 1947; reissued Woodbury, N.Y.: American Institute of Physics, 1995), although inaccurate in parts, is a poignant firsthand account by the head of the U.S. mission entering Europe before the war was over to collect information—and snatch scientists— from the German side.

Physics and Beyond: Encounters and Conversations (London: Allen & Unwin, 1971) is Heisenberg's own account of his life and main intellectual branching points. David Cassidy's *Uncertainty: The Life and Science of Werner Heisenberg* (Basingstoke, England: Freeman, 1992) enriches the story. Out of fairness I should mention *Heisenberg's War: The Secret History of the German Bomb,* by Thomas Powers (London: Jonathan Cape, 1993), which takes a view very different to my own; his thesis is seriously questioned in the lengthy review by Richard Peierls (available in his *Atomic Histories,* (New York: Springer-Verlag, 1997), pp. 108-16, as well as in Jeremy Bernstein, ed. *Hitler's Uranium Club*, the review in *Nature, 363* (May 27, 1993), pp. 311-12, and especially the reviews in the *American Historical Review, 99* (1994), pp. 1715-17, and Paul Lawrence Rose's *Heisenberg and the Nazi Atomic Bomb Project: A Study in German Culture* (Berkeley: University of California Press, 1998).

An excellent account of the Norway events is Knut Haukelid's own *Skis Against the Atom* (London: William Kimber, 1954, reissued), while the brief sections in Leo Marks's extremely readable *Between Cyanide and Silk: A Codemaker's Story 1941–1945* (New York: HarperCollins, 1999) on the solidarity of the Norwegians as they trained in London adds to an understanding of their success. On the UK effort, *Operation Freshman: The Rjukan Heavy Water Raid 1942,* by Richard Wiggan (London: William Kimber, 1986) makes good use of transcripts from the war crime trials later held in Norway, and also captures the befuddlement of the tough London kids sent out to this lethal, wintry terrain.

The American decision to use the bomb is surveyed from a conventional military/strategic viewpoint in *Americans at War,* by Stephen E. Ambrose (New York: Berkeley Books, 1998) pp. 125-38; from an administrative viewpoint in *Britain and Atomic Energy 1939–1945,* by Margaret Gowing (London: Macmillan, 1964). Best of all, though, is J. Samuel Walker's *Prompt and Utter Destruction: Truman and the Use of Atomic Bombs Against Japan* (Chapel Hill, N.C.: University of

North Carolina Press, 1997), which emphasizes how much the ill-prepared Truman was pushed and led by his advisors, with their own bureaucratic, geopolitical, and domestic concerns; also how many of the key American military leaders then would have been startled by the later consensus that the bombing was inevitable.

Whether or not the decision was justified, the accounts in Chapter 19 of Richard Rhodes's *The Making of the Atomic Bomb* are a necessary reminder of what the decisions meant on the ground those two summer mornings in August; the almost aphasic resistance of many postwar researchers to discuss any aspect of the morality of their weapons work is a central topic in *The Genocidal Mentality: The Nazi Holocaust and Nuclear Threat,* by Robert Jay Lifton and Eric Markusen (London: Macmillan 1991).

The Universe (Chapters 14–16)

Payne

The richest source here is *Cecilia Payne-Gaposchkin: An Autobiography and Other Recollections,* ed. Katherine Haramundanis (New York: Cambridge University Press, 2nd ed., 1996). See also George Greenstein's reflective essay "The Ladies of Observatory Hill," in his *Portraits of Discovery* (New York: Wiley, 1998). An interesting comparison from a later generation is *Bright Galaxies, Dark Matters* by Vera Rubin (Woodbury, N.Y.: American Institute of Physics, 1997), while George Gamow's dated though highly readable *The Birth and Death of the Sun: Stellar Evolution and Subatomic Energy* (London: Macmillan, 1941) gives a useful impression of solar physics in Payne's time.

Hoyle and Earth

Fred Hoyle is the best writer of any high-level scientist I'm aware of: his autobiography, *Home Is Where the Wind Blows: Chapters from a Cosmologist's Life* (New York: Oxford University Press, 1997), is a pleasure to read. One learns why his generation of youngsters suffered the wettest feet in Yorkshire (previous generations had clogs, which let water drain through, the next generation had boots, which kept water out, but his had cheap boots, which let water in and kept it there). One also learns about Dirac's lecturing style, Eddington's thinking style, the distortions produced by Cambridge's overdifficult exams, the achievements produced by Cambridge's intensely fair scholarships, as well as pointers on nucleosynthesis, RAF versus Royal Navy research styles, academic politics, and the surprising durability of cardboard cars.

For the wider context in which Hoyle worked, again Timothy Ferris's *Coming of Age in the Milky Way* (New York: William Morrow, 1988) is ideal.

Chandrasekhar

Kameshwar C. Wali's *Chandra: A Biography of S. Chandrasekhar* (Chicago: University of Chicago Press, 1992) is an excellent biography, and the sixty pages of transcripts of Wali's conversations with Chandra in the Epilogue are especially recommended. When Chandra describes Fermi ("The fact, of course, was that Fermi was instantly able to bring to bear, on any physical problem . . . his profound and deep feeling for physical laws. . . . [The] motions of interstellar clouds with magnetic lines of force threading through them reminded him of the vibrations of a crystal lattice; and the gravitational instability of a spiral arm of a galaxy suggested to him the instability of a plasma and led him to consider its stabilization by [a] . . . magnetic field."), he's also describing himself: giving us a glimpse of what it might be like to view the world through such a powerful, interlinking mind. I'd also recommend Chandra's own book of essays, *Truth and Beauty: Aesthetics and Motivations in Science* (Chicago: University of Chicago Press, 1987).

For further topics in astrophysics there are an abundance of fine texts. *The Five Ages of the Universe: Inside the Physics of Eternity,* by Fred Adams and Greg Laughlin (New York: Free Press, 1999), is especially good, covering the story from the earliest moments to a very, very distant future. Stephen Hawking's collection *Black Holes and Baby Universes* (New York: Bantam, 1993) is entertaining and wryly thoughtful; while for the reader who relishes popular science books on the universe but finds they're beginning to blur, I'd strongly suggest stepping back and working through a crisp introductory text such as *The Dynamic Universe: An Introduction to Astronomy,* by Theodore P. Snow (St. Paul: West Publishing Company, several editions).

General Relativity (Epilogue)

The best introduction I'm aware of is also one of the most concise. It's Robert M. Wald's *Space, Time, and Gravity: The Theory of the Big Bang and Black Holes.* To go along with that there's Robert Geroch's equally excellent *General Relativity From A to B.* Both Wald and Geroch take a clear geometrical approach, and have numerous picture diagrams carrying the story along through their texts, so the nonscientist will find them as easy as reading a book on architectural design—only here the design is that of our universe.

Black Holes and Time Warps: Einstein's Outrageous Legacy, by Kip Thorne (New York: Norton, 1994) is much longer, and sometimes loses the thread in its gushing biographical backgrounds. But much of it is vivid, and Thorne, as much as Wald and Geroch, has been a leader in the field of general relativity for decades. For a thoughtful account of the 1919 eclipse expeditions—and Eddington's true motivations—don't miss Chapter 6 of Chandrasekhar's *Truth and Beauty: Aesthetics and Motivations in Science.*

$$E=mc^2$$

I couldn't have written this book on my own. A lot of it developed out of the Intellectual Tool-Kit courses I taught at Oxford, which Roger Owen and Ralf Dahrendorf were central to getting started. Avi Shlaim helped nurture that series over the years, and Paul Klemperer made apt comments after one of the creativity lectures, which helped lay the idea for an expansion of the physics aspects of that course.

Once a first draft was done, several friends were kind enough to read the manuscript in its entirety: Betty Sue Flowers, Jonathan Rowson, Matt Hoffman, Tara Lemmey, Eric Grunwald, Peter Kramer, and Caroline Underwood. They gave excellent suggestions, a number of which I even adopted. George Gibson and Jackie Johnson at Walker & Company were even more valiant: repeatedly offering wise comments that improved the book no end. Readers who looked through particular chapters for accuracy, or answered specific questions, included Steven Shapin, Dan van der Vat, Shaun Jones, Bob Wald, Tom Settle, Malcolm Parkes, Ian Kogan, David Knight, Winston Scott, and Frank James. None of them, of course, are responsible for any errors that remain.

319

Two individuals gave especially important aid. In a series of long, flowing phone talks Doug Borden helped me see how the final visions of the "energy" and "mass" chapters could be best developed. Gabrielle Walker, the most eloquent of friends, talked me through all aspects of the book, opening up a world of honest writing in conversations that sparked across months of fine dinners. In one particularly memorable late-night stroll through St. James's Park, she explained how the quietly widening chorale of the St. Matthew's Passion showed the way to escape from strict chronology after the equation's story reached 1945. The book would have collapsed after chapter thirteen without that.

For a long time I was perplexed about what level of explanation would be best in the main chapters. Peter Kramer, especially, was persuasive in his observation that I needed to give the results of the equations, without slighting the explanation of why the equation holds true. To do this, I put an indispensable core of explanation in the main text, a little more in the notes at the end, then even more—and especially anything that involves mathematics—on the Web site, davidbodanis.com. I like the idea that a book is no longer a single defined object, limited by the technology of paper and glue and stitching. To keep the Web site from being only for technical types, I also included there some reminiscences of boyhood in Chicago (which with only a slight twist lead to an explanation of how space and time slosh into each other). There are also insights from William Blake, samples of Einstein's voice, links to the courses I offer on the equation, a look at why simple art forms such as equations are so often true, and other odds and ends.

The newly finished British Library was an excellent place to research all this: It's one of the great libraries of the world, and possibly the last, pyramid-like homage to the pre-Internet era. Many of the Library's science journals were still in the old Southampton Row reading rooms, where interior design and coffee facilities were not quite at the same level, but the photostats of original patent applications on the wall (Whittle's jet engine, the paperclip, the thermos flask, the Wright brothers' wing-warping) made up for a lot of that.

The University College science library in London was also useful, and even though the physical plant is now showing the effects of years of underfunding, the staff do an excellent job of trying to shore up the gaps. The London Library on St. James's Square doesn't suffer those funding problems and is a strong reason for living in this city. It's an early Victorian institution that still works: There are about a million books, on open shelves, including many early editions. I became used to reading texts that would refer to some earlier biographer's hard-to-obtain work, which could usually be conveniently found, albeit under a light sprinkling of dust, just an arm's length further along the shelf.

There was an added benefit there, since for Faraday and Maxwell and the like I could scoop up armfuls of their works or letters, and head outside to one of the benches under the oaks in the center of St. James's Square. It was a fitting location. To one side was the red-brick building which had housed Eisenhower's SHAEF headquarters in 1944, when the fears of a German atomic bomb were near their peak; behind me was the plaque to Ada, Countess Lovelace, the nineteenth-century predecessor of computer programmers, who experienced many of the ups and downs a woman's career in science was likely to take. Walking to a sushi bar for lunch took me past one of Newton's homes on Jermyn Street; when I finally settled for lunch I was right across from the great hall where the news confirming Einstein's general relativity predictions was released.

Most of the actual writing was done when my wife, Karen, was making a transition from being a distinguished historian, to being a distinguished business consultant. We'd always spent a lot of time with our children, but when she was off in Geneva or Washington or Berlin—although she later helped with draft after draft, giving kind, incisive support—I had even more of the day with them. This meant writing time was often broken up. But curiously the text proceeded faster than before.

What happened, I think, was that by really getting into the time with the kids, I was forced to have the breaks that authors rarely allow themselves. Strolling to school we'd get down on our bellies to observe ants in the grass, or we'd stop and chat with the men drilling the streets, who almost always had younger brothers and sisters, or kids of their own, and so were only too happy to rest and explain how their tools worked to the fascinated three and five year olds. There'd also be wall walking and "secret spy," long lunchtimes and afternoons. There were times when I was grumpily distracted (sorry guys), but mostly I looked forward to our hours together, and the wondrous refreshment that very young, very curious minds provide (thanks guys).

When it finally did get too late for more, and two exhausted youngsters were asleep in their bunk beds, I'd settle into a big chair in their room (it felt a lot friendlier there than being in my study), with notes and bound volumes spread out, and then I'd gladly return for hour after hour to this book, as the sky darkened and the London streets went quiet outside. A few times—the writing racing along; my coffee long since cold—I'd realize I'd gone the whole night through;

most notably once while writing about the chemistry of the sun, as the roaring sphere of that star—powered by thermonuclear blasts in accord with $E=mc^2$—began to lift from behind the Earth, somewhere far beyond the Thames estuary; lifting, rolling, to embrace our lives.

I loved writing this book.

E=mc²

Academy of Sciences
(France), 31, 32, 33, 34, 57,
61
Accelerators, 52
Air, 29
different gases in, 30
Allison, Samuel, 131
Alpha Centauri, 183
Americium, 193
Andromeda galaxy, 201
Annalen der Physik, 7
Anti-Semitism, 217
Arouet, François-Marie
see Voltaire (François-
Marie Arouet)
Astronomy, 42–43
Astrophysicists, 180, 201
Atom, ix, 93–99, 113
empty, 94–95, 99, 125
opened, 111, 112
Atomic bomb, 163, 191, 217
Einstein and, 130, 217, 218
energy from, 77
first, 113
used on Japan, 160–62

work on, in Germany, ix,
124–33, 134–35, 141,
151–52, 158–59
Atomic bomb explosion,
163–69, 168f, 173, 183,
189
Atomic bomb program
U.S., 119, 143–53, 157–58,
160–62, 187–88
Atomic reaction, 127–28,
159

Barium, 106, 107
Barnard, George, 16
Belgian Congo, 129
Bell, Alexander Graham, 15
Berkeley, California, 111, 131
Berlin Auer company, 96,
129
Bernoulli, 62
Besso, Michele, 7, 78, 82,
88–89
Bethe, Hans, 152, 182
Black holes, ix, 199,
200–202

Bohr, Niels, 109, 121, 152–53
  institute, 107, 112, 153
Bombs, 118, 119
  see also Atomic bomb
Brahe, Tycho, 39
Breathing, 35
Breteuil, Emilie de
  see du Châtelet, Emilie
Briggs, Lyman J., 119, 120,
  131, 133, 161, 215
British
  in atomic bomb
    program, 151
  attack on heavy water
    plant in Norway,
    136–41
British intelligence, 134, 153
Brun, Jacques de, 58
Burlington House, 213, 215
Byrnes, Jimmy, 161–62

c (symbol), viii, 52, 79
  bridge connecting energy
    and mass, 69
  for celeritas, 37–54
  squaring, 54
c^2, 69, 77, 99, 112–13, 166,
  183, 194
  bridge connecting energy
    and mass, 111
C-14, 194
Calculus, 84
Cambridge, 175, 176, 197
Cambridge Solar Physics
  Observatory, 175, 181
Cancer, 76
Candide (Voltaire), 64
Carbon, 109, 184, 188, 189,
  190–91
  unstable varieties of,
    193–94

Carbon dating, 194
Casimir, Hendrik, 135
Cassini, Jean–Dominique,
  39–43, 44
Cavity magnetron, 131
Celeritas, 37–54
Centimeters, 53
CERN, 52
Chadwick, James, ix, 96–97,
  127, 157
Chain reaction, 159, 165,
  166, 167
Chalk River, British
  research group at, 150,
  187
Chandrasekhar,
  Subrahmanyan (Chan-
  dra), x, 197–201, 198f,
  209, 213
Chaos theory, 78
Chemical energy, 19, 20, 26
Circle, 15, 16, 45
Cirey, scientific research at,
  60–62, 64, 65, 66
Cleveland, Grover, 119
Cold War, 192
Collisions, energy lost in,
  62–63, 64
Computers, 178–79
Concentration camps, 129,
  135, 139
Conservation laws, 20,
  53–54, 61
Conservation of energy, 20,
  21–22, 35, 53, 67
Conservation of mass, 35,
  53, 61
Conversion factors, 52, 53,
  54, 69
Crouch, Henry, 215–16
Crystalline model, 84

Curie, Marie, 75–76, 80, 92, 124
Cyclotron, 153
Cygnus the Swan, 200
Czechoslovakia, 124, 129

D Day, 152
Danton, 34
Darwin, Charles, 20, 174
Davy, Sir Humphry, 12–13, 14, 14f
  and Faraday, 13, 14, 17–18, 46
Degenhart, Joseph, 5
Delivery weapons, advanced, 152
Denmark, 40, 43
Descartes, René, 58
Deuterium, 126
Dickens, Charles, 18
Dirac, Paul, 186
Domenico, Giovanni
  see Cassini, Jean-Dominique
Döpel, Robert, 127, 128
du Châtelet (soldier), 59
du Châtelet, Emilie, x, 58–62, 60f, 63, 64–67
  death of, 66–67
Dyson, Frank, 210, 211, 214

E (symbol), viii, 37, 52
  for energy, 11–22
E=mc², 123, 199, 204, 206
  "adulthood," ix–x, 115–69
  ancestors of, viii, 9–69
  applications of, 173, 174–75, 191–94
  applied to explain fires of sun, 182–83, 195
  in atomic bomb explosion, 165, 166, 169, 218
  in atomic reaction, 128
  and atoms, 95, 96, 98–99
  biography of, vii–x
  birth of, viii, 8
  coming of age, ix
  in creation of Earth, 185, 190, 191
  effects of, 106
  Einstein and, 73–92
  in end of world, 197, 202–3
  energy promised by, 100, 110–11, 112–13, 125, 130, 141
  extends through universe, ix, 173, 174–75, 183, 187
  in implosion, 189
  in plutonium explosion, 150
  in race to build atomic bomb, 133, 141, 154
E=mc¹, 65
E=mv², 65
Earth, 80, 174, 183, 206
  creating, 184–94
  end of, 196–97, 201–2
  hydrogen atoms, 182
  material substances interconnected on, 27, 35
  movement of, 41–42
Eclipse, 207, 208, 209
  expeditions, 208, 209–12
Eddington, Arthur, 176–77, 181, 187, 200, 201, 209–14, 211f
Egyptian symbols, 24

Ehrenfest, 212
Einstein, Albert, 3–8, 6f, 15,
    36, 37, 91f, 103, 108, 119f,
    120, 121, 139, 159, 173,
    200, 219f
  accomplishment of, 85,
    204–19
  and atomic bomb, 130,
    217, 218
  career, 5–6, 7, 90
  celebrity status, 214–15,
    216–17
  children, of, 90
  disavowing, 123
  and eclipse expedition,
    212
  equation, viii, ix, 26, 44,
    53, 73–92, 110, 112, 137,
    183, 194, 203
  equation confirmed, 213,
    217
  letters to Roosevelt,
    117–19, 123, 129, 130,
    204, 217
  light waves, 48–49, 51
  mc^2 in, 69
  on Meitner, 100
  papers, 4, 7–8, 198
  patent office job, 3–5, 6–7,
    48, 78, 85, 87, 90, 174,
    205
  physics articles
    published, 6
  powers of, fading, 218–19
  reasons for success of,
    85–89
  as student, 5, 21, 22, 48,
    178
  theory of, confirmed,
    213–14, 215, 217
  theory of relativity, 7–8

Einstein, Hans Albert, 74
Einstein, Hermann, 4
Einstein, Maja, 6, 87, 218
Einstein, Mileva, 4, 5–6,
    89–90, 91f
Einstein family, 86, 87, 88,
    159
Eisenhower, Dwight D.,
    152, 161
Electric engine, 16
Electricity, 12, 45
  in atoms, 95
  of light wave, 49
  and magnetism, 13, 15–16,
    19, 47, 50
  in nucleus, 109, 110, 111,
    112
  of protons, 165
Electricity-generating sta-
    tions, 192
Electromagnetic event
  area around, filled with
    "field," 45
Electrons, 95, 97, 164
Elements, 184, 187, 188, 191
  created by stars, 189
  $E=mc^2$ in creation of,
    185
  visual signals, 174
  unstable, 190
Energy, 11–22
  amount of, fixed, 19–20,
    22, 54
  c^2 bridge linking mass
    with, 111
  concept of, 17, 18–21, 22,
    27, 30
  different kinds of, linked,
    17, 19, 20, 22
  du Châtelet's work on,
    62–66

and/as mass, 35–36, 37,
    52–54, 69, 74–76, 81–82,
    91–92, 108, 110, 112, 165,
    166, 182, 183, 198–99,
    203, 213
measurement of, 68
as mv^2, 62, 64, 65, 67, 68
promised by $E=mc^2$, 92
promised by $E=mc^2$,
    allowed to emerge,
    100, 110–11, 112–13, 125,
    128, 130, 141
properties of light and,
    51–52, 54
released in atomic bomb
    explosion, 167–68
sources of, 22, 26
space/time curved
    around, 205–7
types change but total
    amount remains same,
    19–20, 36
Energy conservation, 20,
    21–22, 35, 53, 67
Energy force, 20, 21
Energy of a moving object
    equation representing, 55
Engineers
    Germany, 132–33
= sign, viii, 23–26, 55, 69,
    111, 182
Equations, 26
    see also Einstein, equation
Exit signs, 193
Explosions, mathematics
    of, 188

Faraday, Michael, x, 54, 67,
    73, 178
    energy concept, 11–22, 12f,
    27, 30, 35

magnetic force lines,
    190
and Maxwell, 45–47, 48
FBI, 130–31, 160
Federal Institute of Tech-
    nology, Zurich, 21
Fermi, Enrico, 97–98, 99,
    103, 125, 126, 157–58
Feynman, Richard, 147–48,
    150
Fission, 112
Force lines, invisible, 47,
    190
France, 55–56, 57, 66, 192
    National Assembly, 33
    taxation, 32–33
Franck, James, 102
Franklin, Benjamin, 32
French Revolution, 31, 33,
    53
French scientists, 31
Freundlich, Erwin, 207–9,
    211
Frictional heat energy,
    19
Friedrich, Caspar David,
    159
Frisch, Robert, 107–8,
    109–12, 119–20, 152, 182
Frost, Robert, 196

Galaxies, 201–2
Galileo, 38, 39, 44, 47
Geiger, Hans, 94, 96, 152
Geiger counter, ix, 94, 152,
    194
General Farm, 28, 32, 35
General relativity, 205,
    207–8
Geometry of world, 67–68
Germany, 104, 117–33

Germany (*continued*)
  work on atomic bomb, ix,
    124–33, 134–35, 141,
    151–52, 158–59
Global Positioning System
  (GPS), 83–84, 194
God, 21, 27, 47, 74, 80, 86, 214
  creating universe, 19, 30
  proof of existence of, 62,
    63, 64
Goebbels, Joseph, 120
Gravesande, Willem
  Jakob 's, 65, 68
Gravitation, laws of, 6, 85
Gravity, 204, 206
Grossman, Marcel, 5, 205
Groves, Leslie, 144–45, 160,
  162, 188
Guillotin, Dr., 35

Hahn, Otto, 100–107, 101*f*,
  110, 112
Haller, Dr., 5, 7
Harvard
  Payne at, 176, 177, 178–79,
    185, 187
Haukelid, Knut, x, 140, 154,
  155, 156
Heat energy, 19, 26
  in atomic bomb
    explosion, 166–69
Heavy water, 126–27, 128,
  134–35
  ferry carrying, blown up,
    153–57
  Norwegian factory, 135–42,
    153–57
Heavy water reactor, 151
Hebrew symbols, 24
Helium, 182, 184, 185
  as fuel, 187

  squeezed to create
    carbon, 188
Helium ash, 184, 189, 195,
  196
Heisenberg, Werner, x, 120–
  30, 125*f*, 132–33, 134, 135,
  141, 143, 145, 151, 152, 157
  work of, blocked, 158–59
Hess, Kurt, 104
Hevesy, George de, 98–99,
  153
Hilbert, David, 78
Himmler, Heinrich, 122–23,
  130
Himmler, Mrs., 121–22
Hiroshima, 167, 173, 192,
  218
Hitler, Adolf, 103, 121, 217
Hörlein, Heinrich, 104
Household devices
  $E=mc^2$ in, ix
Hoyle, Fred, 185–89, 186*f*,
  188
Hyades cluster, 209–10
Hydrogen, 31, 126, 184, 189,
  191
  in creation of Earth, 190
  in sun, 180–81, 182–83, 187,
    195
Hydrogen bomb, 191
Hydrogen nucleus, 182, 187

I. G. Farben, 135, 141, 153
Implosion, 150, 152, 158,
  188–89
Inches, 53
Infrared scope, 32
Insight, scientific, 49, 184
Institute for Advanced
  Study, Princeton, New
  Jersey, 117, 217

Io, 40–41, 43
Iron, 35, 184, 190
  in sun, 174, 177, 179–80

Japan
  atomic bomb used on,
    160–62, 163–69, 183
Jet drones, 152
Jet prototype, 145
Jews, 88, 121
  Einstein, 86, 217
  expelled from Germany,
    133
  Meitner, 103–4
Joachimsthal, 124, 129
Jones, R. V., 138, 139
Jupiter, 42, 50, 208, 209
  moon Io, 40–41, 43

Kafka, Franz, 90
Kaiser Wilhelm Institute
  for Chemistry, 103, 104
Kaiser Wilhelm Institute
  for Physics, 104, 130
Kaiser Wilhelm Institutes,
  102, 103–4
Kepler, Johannes, 39, 40
Keynes, John Maynard, 217
King, Martin Luther, Jr., 216
Koenig, 61–62

Lake Tinnsjö ferry, 153–57,
  156f, 158, 160
Larsen, Alf, 156
Lavoisier, Antoine-
  Laurent, 28–35, 39, 54,
  55, 79, 178, 208
  insights from, 73
  killed, 34–35
  rust experiment, 29–30, 61
  work on mass, 29–31, 35

Lavoisier, Marie Anne
  Paulze, 28, 29–30, 32
Law of the Conservation of
  Energy, 20
Law of universal
  gravitation, 85
Lawrence, Ernest, 131,
  143–44, 148, 149
League of Nations, 123
Leahy, Admiral, 160
Leibniz, Gottfried, 62,
  63–64, 65, 66
Leipzig, university of, 124,
  126–27, 128, 130, 132, 135,
  138, 140, 145, 159
LeMay, Curtis, 160–61
Light, 74
  constancy of, 83
  inner properties of, 44–45,
    47–48, 49, 50
  is physical process, 50
Light beam
  electricity-magnetism
    leapfrogging in, 47–48,
    49, 50
Light waves, 48–49
Littlewood, 213
Los Alamos, New Mexico,
  research center, 144–45,
  147–48, 149, 151, 152, 157,
  160, 162, 187, 188
Louis XVI, 28, 31, 32

m (symbol), viii, 37, 52
  for mass, 27–36
MacArthur, Douglas,
  160
Mach 1, 44
Magnetic energy, 26
Magnetic force lines, 47,
  190–91

Magnetism
  and electricity, 13, 15–16,
    19, 47, 50
  of light wave, 49
Manchester University, 93
Mandela, Nelson, 216
Manhattan Project, 131,
  145, 187, 191, 218
Marat, Jean-Paul, 32, 33–34
Marshall, George, ix
Mass, 27–36
  amount of, unalterable,
    30–31, 35, 36, 54
  c^2 bridge linking energy
    with, 111
  concept of, 27
  conservation of, 35, 53, 61
  and/as energy, 35–36, 37,
    52–53, 69, 74–76, 81–82,
    91–92, 108, 110, 112, 165,
    166, 182, 183, 198–99,
    203, 213
  multiplying by square of
    velocity, 68–69
  properties of light and,
    51–52, 54
  single connected whole,
    28
  space/time curved
    around, 205–7
  of sun, 195
Mass times c squared
  (mc^2), 68–69
mv^2, 62, 64, 65, 67, 68
Mass times velocity (mv^1),
  62–63, 68
Material substances
  energy hidden in, 54
  measuring shifts in
    weight or size of, 27,
    29–31

Matter, 202
  deeper mechanisms of,
    100
  energy hidden in, 26
  inner structures of, 92
  links between, 27
  made of atoms, 93
  moving among forms but
    not disappearing, 30,
    35
Maupertuis, Pierre-Louis,
  59
Maxwell, James Clerk,
  45–48, 46f
Measurement systems, 53
Measuring apparatus
  (Lavoisier), 29–30, 32–33
Mechanics, 85
Medical devices, ix, 193
Meitner, Lise, 100–112, 101f,
  117, 127, 178, 182
Mendès-France, Pierre,
  34
Mercury, 195
Middle Eastern symbols,
  24
Milky Way, 201
Milne, Edward, 175, 181
Minkowski, Hermann, 205
Mushroom cloud, 168f, 169

Nagasaki, 218
Nazis, 104, 135, 158
Nedelsky, Leo, 146
Neumann, John von, 150,
  164
Neutron(s), 96, 97–98, 99,
  108, 109–10, 111, 112, 128,
  130, 132
  in atomic bomb
    explosion, 164, 165

as probe, 98, 103
slowed, 98, 105, 125–26,
   127–28, 134, 141
Neutron source, 158–59
New York Times, The, 214–15
Newton, Isaac, xi, 27, 31,
   39, 45, 67, 73, 84–85, 148,
   214
   accomplishment of, 84–85
   energy, 62–63, 65
   laws, 6, 56–57
Newtonian science, 62
Norsk Hydro plant, 136*f*
Norway
   heavy water factory, 134–
      42, 153–57
Norwegian Resistance,
   135–36, 153, 155
Norwegian volunteers
   attack on heavy water
      plant, x, 138–42, 154
Nuclear energy, 192
Nuclear physics, 107
Nuclear reaction, 149, 177
Nuclear submarines, 192
Nucleus(ei), 95, 96, 97, 99,
   112, 124–25
   in atomic bomb explo-
      sion, 164–65, 166
   cracking apart, 113
   energy to tear apart, 108
   gluing strong force, 109
   hydrogen, 182, 187
   model of, 108
   probe to enter, 98, 103
   uranium, 111, 128, 174
   what happens inside,
      100

Object(s), receding, 82
Oliphant, Mark, 131

Oppenheimer, J. Robert,
   145–52, 146*f*, 152, 157–58,
   159–60, 164, 182, 188, 218
   and use of atomic bomb,
      161, 162
Ostwald, Wilhelm, 3–4
Oxygen, 30, 31, 184, 188, 190,
   191
Oxygen molecules, 35

Paris, 32–33, 57, 59
Paris Observatory, 39
Paschen, Wilhelm, 128
Pauli, Wolfgang, 121
Paulze, Jacques, 28, 34, 35
Payne, Cecilia, x, 175–81,
   176*f*, 182, 184, 185, 187, 200
Pearl Harbor, 131
Peenemünde, 152
Pentagon, 148
Periodic table, 103
PET scans, ix, 193
Phoenician symbols, 24
Photons, 167
Physicists, ix, 6, 68, 112, 147
   Allied, 153, 154
   and Einstein, 78, 88
   and E=mc^2 equation, 91
   German, 104
   Jewish, 121, 123
   patronized Faraday, 45
   refugee, 131–32, 143, 148–49
Physics Institute, 101
Physikalische Zeitschrift, 103
Picard, Jean, 39, 40
Planck, Max, 103
Planets, 27, 177
+ sign, 24
Plutonium, 148–50, 151, 152,
   158, 188, 189
Poincaré, Henri, 78–79, 85

Port Libre prison, 34, 39

Princeton, 117, 217

Protons, 96, 99, 108, 110, 111
  in atomic bomb explosion, 165
  cluster together, 109
  decaying, 202
  energy becoming mass in, 52

Quantum mechanics, 120, 159

Radiation, 105, 106

Radiation treatment, 193

Radioactive particles, 191–94

Radioactive substances, 96, 97, 127

Radioactive warheads, 152

Radioactivity, 73, 75–76, 103, 107, 174

Radium, 76, 92
  variants of, 105–6

Reactor, 144, 188

Recorde, Robert, 24–25

Relativity, viii, 82, 194
  Einstein's initial paper on, 73–74, 77–78
Einstein's work in, 117, 123
  term, 84
  and uppermost speed limit, 81–83

Relativity, theory of, 7–8, 53, 79–80
  first, 85
  of Poincaré, 79
  special, 90, 205

Religion
  Einstein and, 86–87
  and science, 62–63

Roemer, Ole, 40, 41–44, 47, 50, 55, 208

Rohan, Chevalier de, 56

Rome, 97, 157

Roosevelt, F. D.
  Einstein's letters to, 117–19, 123, 129, 130, 204, 217

Royal Astronomical Society, 213–14

Royal Institution, 12, 17, 18, 30, 46

Royal Marines, 136

Royal Prussian Observatory, 207, 208

Royal Society, 16, 213–14

Russell, Bertrand, 213

Russell, Henry Norris, 181, 187

Rutherford, Ernest, ix, 93–96, 94f, 99, 102, 107, 131, 137, 145, 147
  treatment of Payne, 176

Sachsenhausen concentration camp, 129

Sandeman, Robert, 14

Sandemanians, 14, 15

Science, 36, 87
  mathematically guided, 80
  and religion, 62–63

Segrè, Emilio, 143–44

Serber, Charlotte, 151

Sexism, 178–79
Shapley, Harlow, 177
Silicon, 184, 188, 190
Simultaneity, 85
Smoke detectors, ix, 193
Solar system, 41–42, 47,
   201
   breakup of, 197
   sun heating, 195
Sonic boom, 196–97
Sörlie, Rolf, 155
Space
   distortion of, 200
   fabric of, 205, 212
Space shuttle example,
   51–52, 53, 74, 81
Space-time
   curved around mass and
      energy, 205–7, 206f
Special relativity, 90, 205
Spectral lines, measuring,
   178–79
Spectroscope, 174
Spectroscope readings,
   179–81
Speed limit, ultimate, 50, 51
   relativity and, 81–83
Speed of light, 36, 37, 51, 52,
   54
   fundamental speed limit
      in universe, 50
   linking mass and energy,
      37, 54, 69, 76
   measuring, 37, 38–44, 50
Speed of sound, 44
Squaring, 55
   of velocity, 67–68
Star core, 198, 199
Stark, Johannes, 121, 122

Starlight, bent, 207, 208,
   210, 212–13
Stars, ix, 77, 177, 182, 184
   dying, 201
   exploding, 198–200
   implosion, 188–89
   using helium as fuel, 187
Stimson, Henry, 161
Stockholm, 105, 106
Straight lines, 15
Strassmann, Fritz, 103, 105,
   106
Strong nuclear force, 108,
   112, 165
Sulfur, 188
Sully, Duc de, 55–56, 57
Sun, 205
   in bending of starlight,
      207, 210, 212–13
   expanding, 195–97
   fires of, 173–83, 184, 195
   warming Earth, 194
Suns, 177

Television sets, ix, 83
Teller, Edward, 150, 152
Temperature, 50
   in atomic bomb, 166–67
   in imploding star, 189
Theoretical physics/physi-
   cists, 48, 73, 181, 204
Third Reich, 123
Thorium, 96, 190
Thought, pure, 80, 176–77
Time
   distortion of, 82, 194, 200
   flow of, 84, 85
   see also Space-time
Tracers, 98–99

Tritium, 193
Truman, Harry, 161
Turin Shroud, 194
² (symbol), viii
Typographical symbols,
    23–25, 55

Uncertainty Principle, 120
Union Minière, 129
United States
    army, 133, 162
    atomic bomb program,
        ix, 131, 133, 143–53,
        157–58, 160–62
Universe, 69, 80, 176–77
    E=mc² in, ix, 173, 174–75,
        184, 187
    end of, 196, 202–3
    energy in, 19
    rational explanation of,
        57–58
    single force in, 20
    views of, 80, 84
University of Berlin, 103
University of Chicago, 2
    01
Uraniborg, observatory of,
    39
Uranium, 77, 92, 117–18, 174,
    190
    in American bomb pro-
        gram, 148–49, 164–65
    in German atomic bomb
        work, 124–25, 134,
        141
    guiding slowed neutrons
        into, 105
    machining, 129
    shape for, 132

Uranium atoms, 141
    in atomic bomb explo-
        sion, 165, 183
    breaking apart, 107–8,
        126, 127–29, 166
Uranium bomb, 77
Uranium nucleus, 107–8,
    109, 110, 174
    exploding, 111, 128
Urey, Harold C., 126, 235,
    149, 157, 158
U235, 149, 157, 158

V–2 missile, 152
Veblen, Thorstein, 85–
    86
Velocity
    multiplying mass by
        square of, 68–69
    squaring of, 67–68
Vemork, Norway, 136, 153,
    154, 155, 156
Venus, 195
Versailles, 58, 61, 64, 66
"Virus House, The," 124,
    130, 138
Volcanoes, 191
Voltaire (François–Marie
    Arouet), 55–57
    and Emilie du Châtelet,
        59–60, 61, 62, 65, 66, 67
    on energy, 62, 63
    supporting Newton's
        ideas, 57

Wandering clubs, 123
Water (H₂O), 98, 126
Watteau, Jean Antoine, 61,
    63

Weapons Bureau
   (Germany), 123, 124,
   141
Weber, Herr, 48
Weber, Max, 79
Wehrmacht, 133, 135, 159
Weimar Republic, 120
Weizmann, Chaim, vii–viii
Western Europe, 21–22
Wigner, Eugene, 131–32

Wilson, Robert, 145
World
   complete system of, 84
   end of, ix, 197, 202–3
World War I, 96, 102, 120,
   132, 208, 209, 210, 214
World War II, 192
World's Fair (St. Louis), 79

X rays, 167